THE

50

MOST EXTREME
PLACES IN OUR
SOLAR SYSTEM

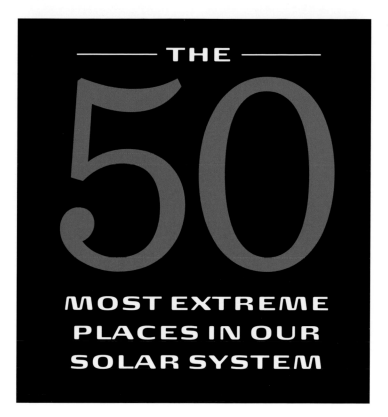

THE

50

MOST EXTREME PLACES IN OUR SOLAR SYSTEM

David Baker and Todd Ratcliff

The Belknap Press of Harvard University Press
London, England • Cambridge, Massachusetts • 2010

Printed in Canada

Library of Congress Cataloging-in-Publication Data
Baker, David, 1967–
  The 50 most extreme places in our solar system / David Baker and Todd Ratcliff.
      p. cm.
  Includes bibliographical references and index.
  ISBN 978-0-674-04998-7 (alk. paper)
  1. Solar system—Miscellanea. 2. Extreme environments. I. Ratcliff, Todd, 1967– II. Title.
III. Title: Fifty most extreme places in our solar system.
  QB502.B345 2010
  559.9—dc22

                                                          2010006126

For Holly, Laney, and Zoe
The best stargazing companions in the Solar System.
— DB

For Max
Who wants his very own copy of Daddy's book.
And for -B-, who helped make it possible.
— TR

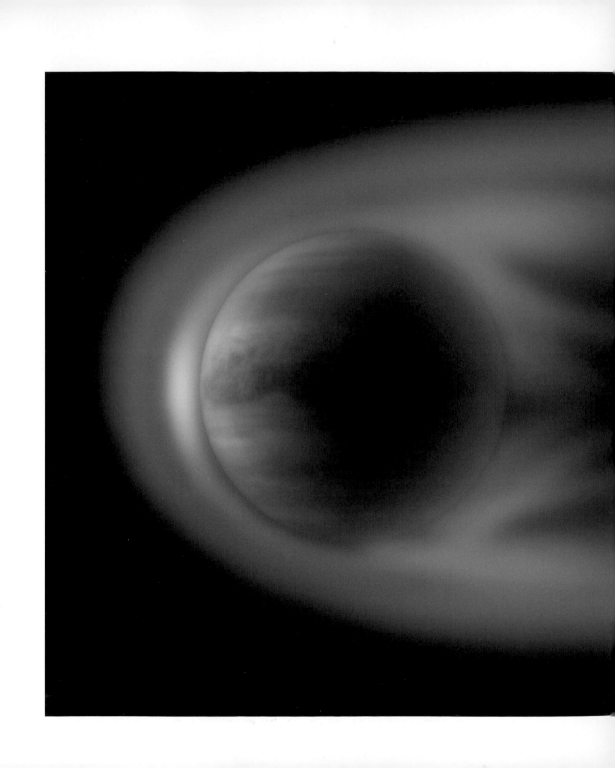

Contents

Preface

It started off as a quirky, fun "What if?" brainstorming session. At the time, one of us (David) was a visiting scientist at NASA's Goddard Space Flight Center doing research on flash flooding. The other (Todd) was, and still is, at JPL using geodynamics and geodesy in support of spacecraft navigation. Down the hall at Goddard, NASA scientists were analyzing data from CAMEX-4, a major scientific mission during which multiple aircraft flew into the dangerous, ill-wind eyewall of Hurricane Erin. "Extreme Weather" was the theme that summer.

During an *important* Goddard-JPL teleconference (we can't remember exactly what got us on the phone that day, but surely it must have been important), we chatted about extreme natural hazards: floods, storms, earthquakes. And then the brainstorming went into high gear. As planetary scientists, we didn't limit ourselves to just extreme weather or even simply to planet Earth. Some intriguing questions emerged: What would we see if we flew through Jupiter's Great Red Spot? Do methane flash floods occur on Titan? Can anything in our Solar System compare to the intense volcanic activity on Io? What *are* the most extreme places in our Solar System?

We decided that someone should write a book about that kind of stuff . . . we would definitely want that book for our bookshelves!

Eighteen months later at the Fall Meeting of the American Geophysical Union, stimulated by hearing about new scientific results and full of good San Francisco food, we talked more about the many extreme places in our Solar System. We quickly came up with at least 60 different extreme things (the list later grew to more than 100). A book was born. But we soon realized that some of the examples were quite superficial—it had begun to look like a laundry list of factoids about the Solar System.

We wanted something more than that. We wanted to delve deeper into the extreme. Certainly, we aimed to capture the "WOW" factor, but at the same time, we wanted to address the most important question in science: Why? We envisaged a book that communicated cool, fascinating, and mysterious ideas—the kind of ideas that inspired us to pursue careers in planetary science in the first place.

We hope that you find an appropriate balance of WOW and serious science in this book. Planetary science is truly interdisciplinary. A mix of the traditional disciplines such as physics, astronomy, chemistry, biology, engineering, and computer science fills the pages of this book. Whether you are a student yearning to explore outer space or a professional scientist dabbling slightly outside of your field of expertise (or anything in between), we hope you find many details about the Solar System to whet your extreme scientific appetite.

To achieve that WOW factor, we have included a lot of amazing images. Each chapter is roughly four pages in length, with half of it text and half of it images. But we don't want this to be a coffee table book . . . we want this book to be *read*! We want pages of the book to be dog-eared, notes penciled in the margins connecting ideas from one chapter to those of another. Science is an active process and this book should be read accordingly. If you don't like to "mess up" your books, you might need two copies—one for your bookshelf and one for your scientific adventures!

There are a number of ways that you can read this book. You can, of course, read it cover to cover, culminating in the final section, "Sum of Extreme Parts," in which the ultimate extremeness of four Solar System bodies is synthesized. This last section definitely will make more sense if you have read the other sections. But we realize that most readers will thumb through the book and first visit those chapters that sound especially appealing—each reader will take her/his own path through the book. To help accommodate this approach (it's one we use quite often), we have tried to make each chapter mostly self-contained. However, it's tough to explain *everything* in four pages! You may occasionally need to refer to the glossary or acronym list, or even turn to a related chapter in the book for more information.

Despite the danger of using metric and English measurement systems together, we have decided to report quantities with multiple units to make the book accessible to a wider audience. For instance, although the preferred unit for temperature in the scientific community is Kelvin, we usually report temperatures in both Celsius and Fahrenheit. These are the units most familiar to people in everyday life. We hope this will help some readers get a sense of the extreme without being sidetracked by unfamiliar units. Just keep in mind that, in practice, it is all too easy to mix up units. The results can be devastating. In 1999, the Mars Climate Orbiter crashed into the Red Planet because two engineering teams used different units, one metric and the other English. It was a $125 million mistake.

If your favorite phenomenon didn't receive a mention in the book, it's our fault. *The* 50 most extreme places actually means 50 things that *we* thought were really cool . . . er, scientifically interesting. The list was debated, dissected, and compiled by a humble panel of two planetary scientists, not by consensus of the entire scientific community. Other scientists might have made different choices. As with potato chips, it's hard to pick just one . . . or even just 50. Incredible new discoveries continue to be made, seemingly every day! We managed to squeeze in two new exciting discoveries—not full chapters but interesting advances in our understanding—just a few days before our final manuscript was due to the publisher.

Some new discoveries we purposely chose not to include. For example, until recently, the coldest place ever *measured by humans* was Neptune's distant moon Triton. Yet in 2009, new measurements found a spot closer to home that is even colder: portions of Earth's Moon that never see sunlight. We believe this "record" will probably soon be broken by a colder place—one we simply haven't measured yet—perhaps detected by NASA's

New Horizons spacecraft as it travels to Pluto and beyond.

We think the extreme places discussed in this book are robustly extreme, that is, they will not be *easily* displaced by new discoveries. Yet as technology progresses and new discoveries are made, we are destined to reexamine our scientific views. Things that we now consider extreme may turn out to be quite commonplace. What if . . . ?

Dive into the Extreme

Get ready to dive into some of the coolest, most incredible things ever discovered here on Earth and throughout the Solar System. As with any extreme adventure, you must prepare yourself mentally for the journey. So let's start by working backward on the title of this book: *The 50 Most Extreme Places in Our Solar System.*

By "Our Solar System," we consider only the region of space dominated by our Sun's gravity. Although there are plenty of interesting things elsewhere in the vast reaches of the universe, we explore in

this book the amazing features of our own celestial neighborhood—planets, dwarf planets, satellites, asteroids, comets, Kuiper Belt objects, Oort Cloud icy bodies, and, of course, the Sun itself—and the processes that make them so remarkable.

The term "Places" is used rather loosely. We may discuss an actual physical *location* such as Olympus Mons, a gigantic volcano that towers above all else on the red planet Mars. We might consider an extraordinary *event* that occurred in the distant past—like the Earth-shaking asteroid impact that annihilated the dinosaurs 65 million years ago. Or it could be a bizarre *phenomenon,* such as rains made of diamond on Uranus and Neptune, not limited to a single location in space or time.

Yet a common thread runs throughout all of these "places": they are unique, curious, out of the ordinary . . . in other words, "Extreme"! They could be the biggest or fastest or the most _____ (fill in the blank). For many extreme places, there is often nothing else like them in the Solar System. Sometimes the extreme examples are just plain awesome. And there are some surprises—a few things that you may consider completely normal are absolutely off the charts!

In this book, we investigate 50 unquestionably extreme places. Trust us, coming up with this list of only 50 wasn't easy. We organize extreme places into basic categories (such as "Rings and Things," "Oceans & Water & Ice—Oh My!" and one of our favorites, "The Wack Pack"), although some topics may, in fact, span multiple categories. We try to convey more than just facts about a particular extreme place; we examine *why* the example is so special and the underlying science behind the extremeness. As often happens in science, there are sometimes just as many unresolved mysteries as answers.

Within each category, extreme places remain unranked (there is no order of importance within each section). Nor do *we* rank the extreme places from 1 to 50 as a whole. That's where you come in. You can vote for your top extreme places at our companion web site, www.ExtremeSolarSystem.com. All we ask is that you consider all 50 chapters before voting—you may find something new, exciting, and unexpected that jets way up on your list.

So strap on your own "Wack Pack" and prepare for a wild read. You may never look at our Solar System in the same way again.

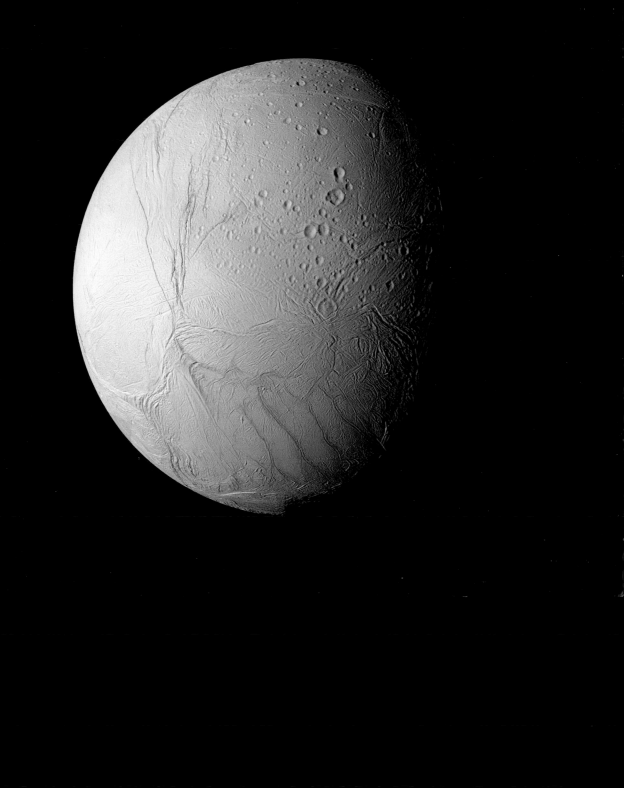

Tallest Mountain—Olympus Mons

Everyone knows that Mount Everest is the biggest mountain around, right? Well, not really. Although Everest may be Earth's *highest* peak—nearly 9 km (5.6 miles) high, as measured from sea level—it is not really our world's tallest. In fact, the *tallest* mountain in the Solar System isn't found on our home planet at all but is an Olympic-sized volcano on Mars.

Earth's *tallest* mountain is Mauna Kea, one of the volcanoes found

Overhead view of Olympus Mons on Mars—the tallest mountain in the Solar System—as seen by the Viking 1 Orbiter. The large summit caldera is 25 km (15 miles) across and was likely caused by collapse after magma drained from the main volcanic vent.

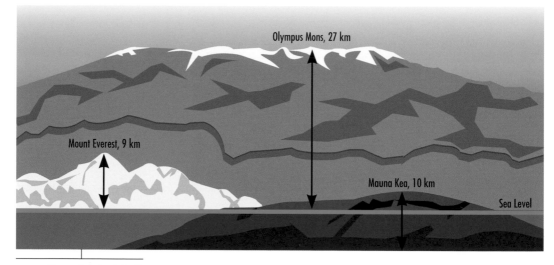

Olympus Mons, 27 km

Mount Everest, 9 km

Mauna Kea, 10 km

Sea Level

Olympus Mons dwarfs Earth's tallest mountains.

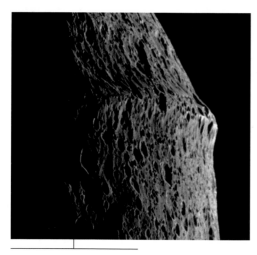

Saturn's moon Iapetus bulges out near the equator, producing a ridge of mountains twice as high as Everest.

on the Big Island of Hawaiʻi. Mauna Kea's total height, measured from its base at the bottom of the Pacific Ocean to its peak high in the Hawaiian sky, is a little over 10 km (6.2 miles). That's a kilometer greater than Everest's base-to-peak measurement!

Many mountains in the Solar System surpass Mount Everest in height. Parts of Maxwell Montes, a mountain range in the northern highlands region of Venus—and the only feature on Venus to be named after a man (James Clerk Maxwell)—edge out Mauna Kea in the height contest at 11 km (6.8 miles). Meanwhile, on Jupiter's fiery moon Io, the peaks of Boösaule Montes reach heights of 18 km (11 miles), about twice as tall as Everest.

The Cassini spacecraft, during recent explorations of Saturn and its moons, discovered a ridge of mountains running nearly a third of the way around the equatorial region of the moon Iapetus. While the origin of this ridge or why it formed on the moon's equator remains unclear, measurements show that parts of the mountain range tower as much as 20 km (12 miles) over the surrounding plains.

It is, however, Olympus Mons on Mars that reigns supreme as the tallest mountain in the Solar System. Located in the equatorial Tharsis region of Mars, Olympus Mons is the tallest of several volcanoes on the Red Planet. The nearby volcanoes, which make up the Tharsis Montes, range in height from 14 to 18 km (9–11 miles), but majestic Olympus Mons surpasses them all. Three times taller than Mount Everest, rising 27 km (17 miles) above the surrounding plains and measuring about 624 km (388 miles) across, Olympus Mons covers about the same area as the U.S. state of Arizona.

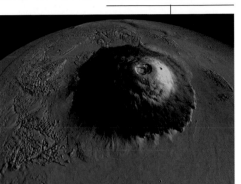

False-color image of Olympus Mons using height data from the Mars Global Surveyor (MGS) spacecraft. The 6-km-high (3.7-mile-high) scarps at the base of the mountain are taller than Africa's famous Mount Kilimanjaro.

Like the islands of Hawai'i, the great martian volcano is thought to have been formed over a "hot spot" where a plume of hot (but not molten) rock rises up to the surface from deep within the interior of the planet. As the hot plume reaches the surface, melting of the rock occurs and volcanoes can be formed.

The great size of Olympus Mons is most likely a consequence of the *lack* of plate tectonics on the Red Planet. Unlike on Mars, Earth's surface is broken up into tectonic plates that move with respect to each other at speeds of a few millimeters per year. At Hawai'i, a hot rising plume impinges upon the bottom of the Pacific plate. Like a conveyor belt moving over a hot flame, the Pacific plate drifts slowly over the plume. Volcanoes form, die out, and form anew as the the plate moves over the hot spot, producing a long chain of volcanoes making up the Hawaiian Islands.

With no moving plates on Mars, Olympus Mons sat above its volcano-forming plume for a much longer time. Although probably long extinct, Olympus Mons was able to grow so large that it now has nearly 100 times the volume of Mauna Loa, Hawai'i's biggest (but not quite tallest) volcano.

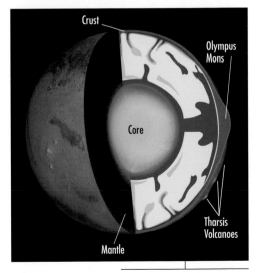

A hot plume rising through the mantle of Mars may be responsible for the largest volcano in the Solar System.

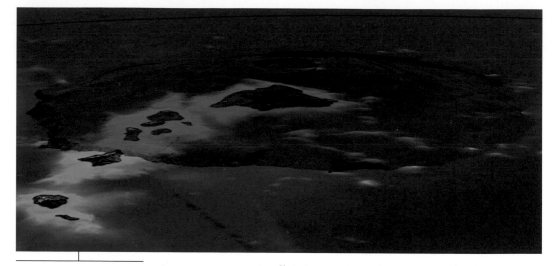

Olympus Mons would completely cover the main islands of the Hawaiian Islands chain if we could drop the monster martian mountain in the Pacific Ocean.

So while the monster volcano of Mars might seem like a tempting target for mountain climbers bored with scaling Mount Everest, they would probably find the trek a bit disappointing. It turns out that scaling Olympus Mons is not so much a technical climb as it is a long walk uphill. With generally mild slopes, it's a climb better suited to hikers than mountaineers. In any case, it's probably a little early to start planning a hiking trip. Considering the eight months required to get to Mars, the several weeks needed for the actual ascent of the mountain, the eight months to get back home, plus the need to take all food, water, and breathable air with you, it's unlikely that travel agencies will be offering this extreme adventure package anytime soon.

Coldest Volcanoes—Cryovolcanoes

When you're hot, you're hot. When you're not, you're not. Cryovolcanoes are definitely *not* . . . hot that is. We're accustomed to thinking of volcanoes as fiery mountains spewing fountains of molten lava and belching noxious clouds of eruptive gases and ash. That's not entirely the case for cryovolcanoes, however. The prefix *cryo* (from the ancient Greek *kryos*) means cold or frozen. Cryovolcanoes are just what they sound like: cold/frozen volcanoes. What these volcanoes spew are chilly plumes of ice and gas at temperatures well below what we consider freezing.

We have nothing like cryovolcanoes here on Earth. We do have *subglacial* volcanoes, mostly in places such as Antarctica and Iceland, but these volcanoes are simply regular, hot volcanoes that lie beneath an overlying ice sheet. So while you might need a heavy winter coat to visit a subglacial volcano, the actual volcano itself is like the others we are familiar with—formed by molten rock.

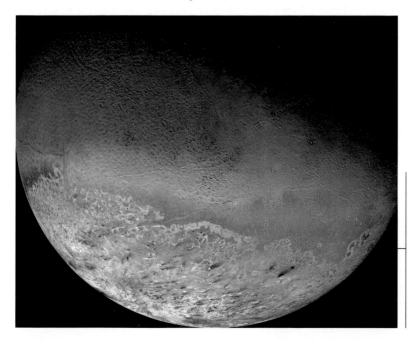

Voyager 2 mosaic of Neptune's Triton. The south polar pinkish deposits indicate an ice cap of older methane ice, while the blue-green band around the equatorial region is probably fresh nitrogen frost. The dark streaks are ice and dust deposited by suspected cryovolcanic geysers. The geysers' plumes can rise as high as 8 km (5 miles) into Triton's atmosphere before being blown by winds and deposited on the surface.

A Dr. Seuss–like ice tower has formed atop a fumarole on Antarctica's most active volcano, Mount Erebus.

A winter phenomenon found along the shorelines of the U.S. Great Lakes is often also called an *ice volcano*. However, the process responsible for these "volcanoes" is definitely *not* cryovolcanism. These ice volcanoes are instead caused by wind-driven waves on the lake. These waves force water beneath ice shelves that extend from the shore. As the waves travel under the shelf, water is forced up through cracks in the ice. This "erupted" water quickly freezes on the surface of the ice. A "volcano" is slowly built, growing taller as additional waves push more water up through the growing vent. Cold and icy . . . but not a volcano.

The coldest true volcano on Earth is Ol Doinyo Lengai in Tanzania. Its name means "Mountain of God" in the Maasai language and it is unique because it is the only active volcano on Earth that doesn't erupt silicate lava. Lengai, instead, spews out a carbonate-rich lava— comparable in composition to laundry soap—that has the consistency of olive oil. Whereas basaltic (silicate-rich) lavas are often erupted at temperatures of 1100°C (2000°F) or more, the natrocarbonatite lava of Lengai is only about 510°C (950°F). Not exactly chilly, is it?

The first evidence for actual cold-as-ice cryovolcanism came from Voyager 2's 1989 flyby of Neptune's moon Triton. Giant geyserlike plumes of nitrogen gas and dust particles were observed rising near the south pole and being blown downwind.

Voyager 2 found the surface of Triton to be incredibly cold at –235°C (–391°F). Triton seems to be mostly covered with frozen nitrogen; it is the explosive eruption of this nitrogen ice that is thought to be the source of the cryovolcanic geysers. Just why does the icy crust burst forth with plumes of gas and dust? One explanation is that the nitrogen-ice layer acts like the panes of a greenhouse (or the glass windows of an automobile on a sunny day). Although relatively weak at Triton's distance from the Sun, sunlight passes through the nitrogen ice and warms pockets of darker material trapped beneath. As the subsurface temperature rises, nitrogen vaporizes and rapidly forces its way to the surface. Unlike the interior of your car, only a small amount of heating is needed for Triton's nitrogen to erupt explosively—a rise of 2–4°C (3.6–7.2°F) is all it takes.

Alternately, warm ice may be brought up from below by convection in the nitrogen-ice layer. Or heat may be released as nitrogen ice changes from one crystalline state to another. Perhaps all three processes are at work. Regardless of how it happens, the gases that these geysers spew out may be as chilly as –231°C (–384°F). Water freezes at a relatively balmy 0°C (32°F), so, by comparison, these cryovolcanic geysers are extremely cold indeed!

Triton isn't the only example of eruptive cryovolcanism in the Solar System. The Cassini spacecraft recently found signs of cryovolcanic activity on Saturn's moons Titan and Enceladus. On Titan, Ganesa Macula and Tortola Facula are two likely cryovolcanoes. While Tortola is only 30 km (19 miles) in diameter, Ganesa is a 180-km (112-mile) dome that resembles the pancake domes found on Venus or perhaps shield volcanoes on Earth. And although it is likely warmer than the –180°C (–292°F) of Titan's frigid surface, the erupted cryolava

Lapping waves form an "ice volcano" along the shore of Lake Superior.

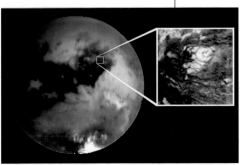

False-color mosaic of Saturn's moon Titan taken by the Cassini VIMS instrument. Inset is a close-up of Tortola Facula, a possible cryovolcano.

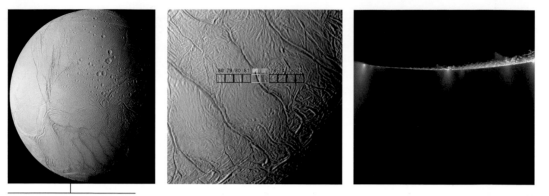

Linear fractures near the south pole of Enceladus (left) have been dubbed the Tiger Stripes. These fractures are considerably warmer than their surroundings. The colored squares in the center image represent temperatures using the Kelvin scale as measured by the Composite Infrared Spectrometer (CIRS) instrument aboard Cassini. The plumes spewing water ice and vapor from the Tiger Stripes (right) are thought to be responsible for creating Saturn's E-ring.

(possibly a slurry containing methane, ammonia, and water ice) is still on the rather chilly side.

The "Tiger Stripe" fractures in the south polar region of Enceladus have been observed by Cassini to be spewing out plumes consisting mostly of water vapor and water ice particles, as well as trace amounts of other gases such as methane, carbon dioxide, and nitrogen. The timing and strength of the plumes appear to coincide with the tidal tugs that Saturn exerts on the icy moon. When Saturn's tidal forces compress the Tiger Stripes, geyser activity drops off. When the tidal pull puts the fractures under tension, the geysers begin to erupt again.

Average surface temperatures of the south polar region of Enceladus are around –188°C (–307°F) or so, but the Tiger Stripes can be nearly 70°C (126°F) warmer. This temperature difference is similar to that of the erupted water of Old Faithful at Yellowstone National Park. The high temperatures and large amount of water vapor in the Tiger Stripe plumes hint at a source of water beneath the surface of Enceladus.

So while Ol Doinyo Lengai is cool for an Earthbound volcano, it still can't match the eruptive cryovolcanism of the outer icy satellites for extreme chill factor. Cryovolcanoes are the c-c-coldest volcanoes in the Solar System.

A Truly "Grand" Canyon—Valles Marineris

Have you ever ridden a mule to the floor of the Grand Canyon in the southwestern United States? Maybe you've taken the 16-hour trip aboard "Chepe" (the Chihuahua-Pacific Railway) through the Barranca del Cobre (Copper Canyon) complex near Chihuahua, Mexico. Or perhaps you've kayaked through the remote and infamously dangerous Yarlung Tsangpo Canyon in eastern Tibet. Maybe instead you've dreamed of descending into Earth's darkest depths—such as the famed Challenger Deep in the Mariana Trench—in a small deep-sea submersible.

What do these massive holes in the ground have in common? They are some of largest crevices on Earth. But in terms of grandeur, they

Valles Marineris, the grandest canyon in our Solar System, was discovered in 1972 by NASA's Mariner 9 spacecraft. This gigantic scar across the face of Mars is a massive complex of canyons that extends nearly one-fifth the distance around the planet. Heavily fractured Noctis Labyrinthus can be seen at the left end of the complex valley system. Melas, Candor, and Ophir chasmata make up the central widened area, and chaotic terrain occurs at the eastern end of the canyon. This image is a mosaic composed of 102 Viking Orbiter images.

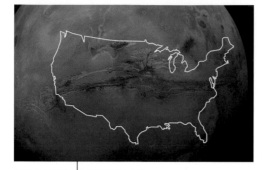

On Earth, the monster martian moat would be large enough to stretch from San Francisco to Washington DC.

don't even come close to the immense gash that cuts deeply across the surface of Mars: Valles Marineris.

Named after Mariner 9—the spacecraft to first detect and photograph the gigantic gorge—Valles Marineris is an enormous fissure just south of the equator of Mars. The extreme canyon complex is more than 4,000 km (2,485 miles) long, 50–100 km (31–62 miles) across, and, in places, over 10 km (6.2 miles) deep. If we could transport Valles Marineris to Earth, the canyon system would extend across the entire United States.

Perhaps the most comparable canyonlike features on our planet, in terms of overall scale, lie deep beneath Earth's seas: ocean trenches. The Peru-Chile Trench off the west coast of South America is 6,000 km (3,700 miles) long, while the Mariana Trench, near Guam, may reach depths as great as 11 km (6.8 miles) below sea level. Although very grand, they are not exactly canyons. Instead of being carved by flowing water, ocean trenches are depressions in the oceanic lithosphere where one tectonic plate is being subducted beneath another.

Trenches can also be quite asymmetric. On the continental side of the Peru-Chile Trench, the Andes Mountains reach an average height of 4 km (2.5 miles) above sea level. Since the deepest part of the trench lies 8 km (5 miles) below sea level, the continental side of

The Grand Canyon is barely a ditch compared to vast Valles Marineris. At certain spots, the walls of the martian canyon are nearly as high as Mount Everest. Earth's ocean trenches have this type of scale on only one side.

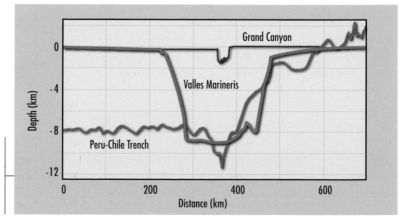

the trench exhibits 12 km (7.5 miles) of total relief! The oceanic side of the trench, however, does not display the same extreme topography. The trench is only about 3.5 km (2.2 miles) deeper than the abyssal plains (the flat part of the ocean floor) to the west.

Similar asymmetry occurs for the Mariana Trench, even though the subducting plate and the overlying plate are both oceanic lithosphere. Measured from the surrounding undeflected seafloor, then, these trenches are generally only 4–5 km (2.5–3.1 miles) deep and less than 50 km (31 miles) wide. Their asymmetry prevents these ocean trenches from being the grandest canyons in the Solar System. They are only half as grand as Valles Marineris!

The great martian gully is thought by planetary geologists to be a large rift valley, similar to the East African Rift on Earth, where the African Plate is in the process of pulling apart to form two new plates. Although Mars is a one-plate planet—it lacks moving plates and thus plate tectonics—it still experiences tectonic motions. As with the East African Rift, the tectonic process responsible for creating the vast canyon system on Mars is uplift.

In the case of Valles Marineris, it is the uplift associated with the formation of the Tharsis Bulge. As hot material from the mantle rises beneath Tharsis, rocks melt, volcanoes percolate, and the entire region is buoyed up. The surrounding crust stretches and cracks. Faults, fissures, and fractures form. Valles Marineris opens—succumbing to the slow, inexorable strain of regional uplift.

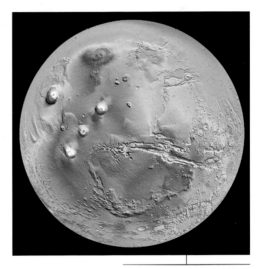

This false-color topographic map from the Mars Orbiter Laser Altimeter (MOLA) shows the relationship between Valles Marineris and the Tharsis Rise. Blues and greens represent low-lying regions; reds and whites are higher regions. The blue-colored floor of Valles Marineris cuts eastward through Tharsis (the large, red, circular region) just south of the equator.

Although not *created* by water, erosion by water likely played a significant role in the evolution of the vast network of canyons. Release of groundwater leads to widening of the rift by undermining canyon walls. As the crust is torn apart in response to the uplift, water and melting subsurface ice are freed to rush outward into the forming crevice. Suddenly bereft of support, the ground collapses and the debris is carried away by the ensuing flood.

While it is likely that most flooding occurred in stages, there are

A perspective view of Valles Marineris looking toward Melas Chasma, parts of which were perhaps once filled with water. This image is assembled from a combination of high-resolution infrared images from the Thermal Emission Imaging System (THEMIS) multiband camera aboard NASA's Mars Odyssey spacecraft and from MOLA altitude measurements from Mars Global Surveyor (MGS).

signs of at least one large catastrophic outburst, as evidenced by the jumbled and collapsed terrain at the eastern end of the canyon. Was it simply large-scale evacuation of groundwater and melting permafrost, or could the flood have resulted from the breach of an ancient lake? Deep chasmata such as Melas Chasma (a broad canyon near the center of the Valles Marineris canyon network) show signs of relatively smooth and flat, layered deposits. These interior layered deposits could be due to falling volcanic ash—or they might have formed from a buildup of sediments in a large body of water. Depths of at least 1 km (0.62 mile) are estimated for such paleo-lakes. A breach, similar to a glacial lake outburst on Earth, would send massive amounts of water gushing toward the outflow channels.

One of the mysteries still puzzling planetary scientists about the Red Planet: if Mars used to be warm and wet, where did the water go? Perhaps the grandest canyon in the Solar System holds some of the clues.

Movers and Shakers—Plate Tectonics

It's a day like any other. You're going about your usual routine. Suddenly, you hear a low rumble . . . feel a jolt . . . and then the shaking starts! *EARTHQUAKE!*

If you're lucky, it's over before you fully realize what's happening. If you're not lucky . . . well, things could be very bad indeed.

Earthquake-related deaths can number in the hundreds of thousands and the destruction of property from major quakes can cost billions of dollars. Fortunately, the vast majority of earthquakes are small. The U.S. Geological Survey (USGS) estimates that several million earthquakes occur each year, but only about 20,000–30,000 are actually detected by the National Earthquake Information Center (NEIC). The others are too small to be measured or occur in remote regions with poor seismometer coverage.

What causes the Earth to tremble so? In ancient times you would have lamented the displeasure of the gods. Today, you could blame it on the accumulation of stress or discuss crustal failure. But ultimately, it comes down to plate tectonics.

In the 1960s, the emerging theory of plate tectonics shook the

The surface of the Earth is made up of several fairly rigid, relatively thin tectonic plates that move relative to each other. On this topographic map, plate boundaries are depicted as thin red lines. White circles represent all significant earthquakes since 2150 BCE. Volcanoes are depicted as red triangles. Notice that both earthquakes and volcanoes tend to cluster around plate boundaries (for example, the "Ring of Fire" that encircles the Pacific basin).

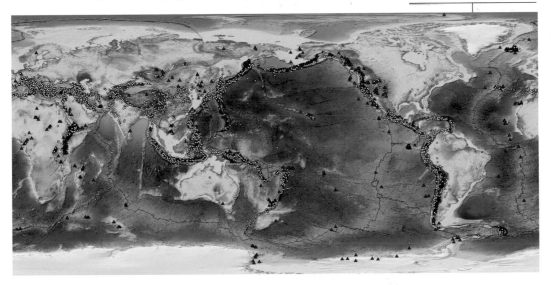

foundations of traditional geophysics. The new theory readily explained seemingly disparate observations: continental drift, midocean ridges, the long island chain of Hawai'i, the Himalaya Mountains, similar fossil records found in both warm Africa and cold Antarctica, locations of volcanoes, magnetic stripes on the ocean floor, and, yes, earthquakes. It was a paradigm shift in scientific thinking comparable to the Copernican revolution in 16th-century astronomy (in which the Sun, rather than Earth, was recognized as the center of the Solar System).

It all starts with the Earth trying to cool off. In addition to heat left over from our planet's formation, Earth's interior is continually warmed from within by the intense radioactive decay of uranium, thorium, and potassium. Convection in the mantle is the way Earth rids itself of all this excess heat.

The viscoelastic nature of mantle rock lets the mantle respond to this heating by "flowing" over geologic timescales (hundreds of thousands of years). This behavior can be illustrated with Silly Putty. By pushing slowly and firmly, you can smoosh and shape the Silly Putty fairly easily. Form the putty into a ball and throw it sharply against a hard surface, however, and the previously soft, pliable putty bounces

The transfer of heat via mantle convection drives plate tectonics on Earth. New crust is created at spreading centers, and old crust is returned to the mantle at subduction zones. Plates move a few millimeters per year — about the rate that your fingernails grow.

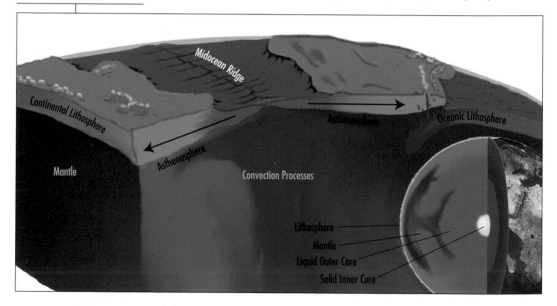

Continental Lithosphere

Midocean Ridge

Asthenosphere

Oceanic Lithosphere

Asthenosphere

Mantle

Convection Processes

Lithosphere
Mantle
Liquid Outer Core
Solid Inner Core

like a hard rubber ball. Over short, fast timescales it behaves as an elastic solid, but over long, slow timescales it behaves like a viscous fluid: viscoelasticity.

Just as hot air is less dense than cooler air—it's what allows a hot air balloon to float through the sky—so too is rock warmed by heating in the deep interior less dense than the cooler rock around it. The warm, less-dense rock "flows" upward toward the surface, where it gives off its heat. As the rock cools, it becomes more dense and eventually sinks back into the interior. This transfer of heat by moving blobs of hot and cold rock is mantle convection.

In contrast to the viscous mantle, Earth's outer layer—the lithosphere (*litho* means rocky)—is quite rigid and is fractured into eight major plates and numerous minor plates. The plates, which carry both oceanic crust and continental crust, float on a low-viscosity zone at the base of the upper mantle known as the asthenosphere. The asthenosphere "lubricates" the bottom of the lithosphere and allows the plates to move relative to one another. The plates continually bump and grind as the "fluid" lower mantle slowly churns beneath them.

This moving drives much of the shaking . . . and volcanism and mountain building and ocean basin formation and . . . well, you get the picture. Whether it's the massive Pacific Plate or the tiny Juan

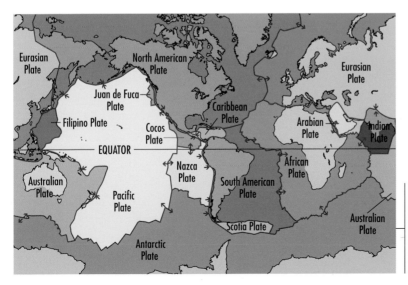

Red arrows illustrate the relative directions of motion between the major tectonic plates. Note the divergence in the mid-Atlantic, the convergence between India and Eurasia, and the strike-slip motion along the western United States.

On Earth, as molten rock solidifies, iron grains within the magma align with Earth's magnetic field. Earth's magnetic field has flipped directions many times in the past, so the cooling oceanic crust records the magnetic field orientation at the time of its formation. As the seafloor spreads apart, magnetic "stripes" move away from the oceanic ridge.

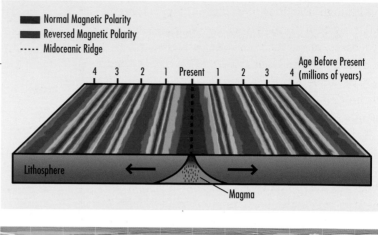

Mars Global Surveyor (MGS) measured magnetic stripes on Mars that are reminiscent of the lineaments that record the seafloor spreading and geomagnetic reversal on Earth.

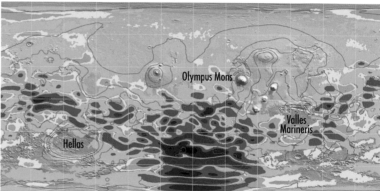

de Fuca Plate, most of the action occurs at the plate boundaries. New crust is created as plates move apart (such as at the Mid-Atlantic Ridge that bisects the Atlantic Ocean). Deep oceanic trenches (such as Challenger Deep in the western Pacific, the deepest spot on the ocean floor) are found at subduction zones, where plates converge and dense oceanic lithosphere dives back in to the mantle.

When continents collide, neither continental plate can subduct. In this case, either the plates are forced upward to build mountains (as with the collision of India and Asia that forms the Himalayan mountain range) or they grind past each other along a strike-slip transform fault (like California's famous earthquake maker, the San Andreas Fault).

Although Earth is not the only place in the Solar System to experience volcanism, tectonic processes, or earthqua . . . , er, seismic activity, it is, as far as we've been able to tell, the only planet to experience *global* plate tectonics. No other body in the Solar System seems to possess the active, mobile surface of our own home planet.

This lack of plate tectonics throughout the rest of the Solar System may not have always been the case. One of the defining observations that led to the theory of Earth's plate tectonics was the measurement of magnetic lineaments on the seafloor. As new seafloor is created at midocean ridges, alternating stripes of remnant magnetization are locked into cooling lithosphere, preserving a record of Earth's flip-flopping geomagnetic reversals.

Mars Global Surveyor (MGS) has detected similar magnetic striping on the surface of Mars. Although these stripes suggest that plate tectonics *may* have operated sometime in Mars's history, the magnetization is seen in only very old rocks, implying that any plate tectonic motions have long since died out. Of course, not all planetary scientists are convinced that the martian stripes are indicative of plate tectonics at all.

Perhaps more intriguing, local "ice" tectonics have recently been observed on Saturn's tiny moon Enceladus. This icy satellite has a series of four major fractures, called Tiger Stripes, located near the south pole. Images from the Cassini spacecraft during close flybys of the moon in 2008 suggest conveyor-belt-like motion away from the Tiger Stripes. Although strike-slip-ingly similar to Earth's seafloor spreading, the ice spreading on Enceladus seems to occur only on one side of the fractures. Local tides (caused by Saturn's strong gravitational pull) in a thin ocean beneath the ice probably drive tectonics near the Tiger Stripes. The rest of the tiny moon's surface appears to be relatively inactive.

Why does large-scale plate tectonics occur only on Earth? It could

---

Plate Tectonics in Your Kitchen

- Add 3 liters (about ¾ gallon) of milk to a large pot over medium heat.
- Carefully add powdered hot chocolate mix to the top of the milk until you have a smooth layer of powder 1–2 cm (about ½ inch) thick—you'll need a large container of cocoa mix.
- Heat the milk to about 87°C (190°F)—it takes about 20–25 minutes—and wait for plate tectonics to begin! Look for faulting, plate motion, subduction, and eruptive volcanism.
- Stir thoroughly and drink your experiment with a dozen or so of your closest friends.

Credit: Professor Dan Davis, Stonybrook University

A comparison of seafloor spreading on Earth (left) and Tiger Stripe spreading on Enceladus (right). Spreading ridges often have a zigzag pattern with strike-slip transform faults where some sections of the ridge are spreading faster than others.

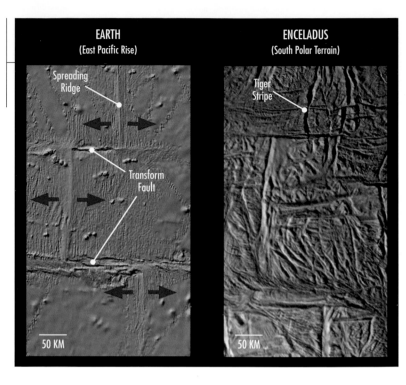

be due to our planet's relatively large size. The larger the planet, the more heat that is trying to escape from the interior and thus the more vigorous the convection in the mantle. Since Mars is smaller, the Red Planet cooled sufficiently long ago that plate tectonics halted (if it ever existed there in the first place).

Perhaps plate tectonics needs liquid water to survive. Not only must there be active mantle convection, but the lithosphere must also be able to fracture into individual plates. Water helps weaken the lithosphere so that it can break, reduces the friction between plates as they slide past one another during subduction, and reduces the viscosity of the mantle so that the rock may "flow" more readily.

Or is there some other key factor we've not yet discovered? Planetary scientists will have to keep studying those planetary bodies with plate tectonics—and those without—to find out why Earth is the leading mover and shaker in our Solar System.

Ooh! That'll Leave a Mark—Extreme Impact Craters

If you've ever played paintball, you know that a small round object smacking into you at high speed can be quite painful. Sometimes you are left with red, swollen welts that last for days. When the Solar System plays its own version of interplanetary paintball, the projectiles aren't necessarily small and the marks—impact craters—can last for millennia.

Impacts are ubiquitous throughout the Solar System. In a sense, impacts helped *form* the Solar System. After all, accretion of planetesimals into planets is simply the process of stuff slamming into each other . . . and sticking together. It's no surprise that we see

That's no space station, that's a moon! Saturn's moon Mimas resembles a famous planet-busting space station from the movies. A large impact almost shattered the small icy moon, leaving a massive crater whose central peak is nearly as tall as Mount Everest. Herschel Crater is 130 km (81 miles) across—nearly one-third of Mimas's diameter—and 10 km (6.2 miles) deep.

The Barringer Crater at Meteor Crater, Arizona, is a classic bowl-shaped crater with a raised rim. The crater, over 1 km (0.6 mile) wide and 200 m (660 ft) deep, resulted from an impact containing as much energy as a large thermonuclear weapon.

Victoria Crater, an 800-m (0.5-mile) impact feature on Mars, was explored by the Mars Exploration Rover Opportunity for more than 14 months. The crater shows extensive modification. The rim has become scalloped by landslides. Smaller craters dot the interior as well as the rim. Sand dunes have formed on the crater floor.

impact craters everywhere: on comets, asteroids, icy moons, and terrestrial planets. We've even been lucky enough to observe destructive impacts as they pounded into the gas giant Jupiter.

While impacts themselves are not all that unusual, some of the marks left behind are pretty extreme. The typical bowl-shaped impact crater, like the one found at Meteor Crater, Arizona, is formed in a three-step process. First, the incoming high-speed impactor (also known as a bolide, often a fragment of an asteroid or comet) must survive its fiery trip through the atmosphere (assuming the target body even has an atmosphere). The subsequent supersonic collision with the ground quickly converts the bolide's enormous amount of kinetic energy into heat and intense compressional shock waves—in both the ground and the impactor itself. This superheated initial stage takes only a couple of seconds or less, usually melting or completely vaporizing the impactor in the process.

The next phase is the response of the ground to such rapid compression: excavation. As shock waves are reflected and bounced around in the ground beneath the impact, the material reaches the limit of what it can bear. When its mechanical strength is exceeded, the ground rock (or ice, in the case of icy moons) shatters. Shock wave energy is converted back into kinetic energy—this time of the fractured ground—that blasts fragments out of the developing transient crater. A relatively large crater can be excavated in a matter of minutes. The explosive nature of the excavation stage so closely resembles a bomb going off that the same computer models used to simulate nuclear explosions can also be used to simulate impact cratering.

The final stage, the modification stage, is often all we ever get to see. Excavated material, blasted out just moments earlier, begins to fall back to the ground. Unstable sections of the crater walls collapse

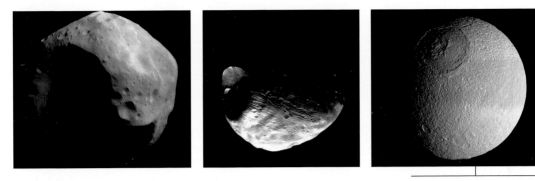

More large craters on small bodies. The small asteroid Mathilde (left) has at least five craters larger than 20 km (12 miles) across—about half of the asteroid's size. Stickney Crater is also nearly half the diameter of the martian moon Phobos (center) at 9 km (6 miles) across. Saturn's icy moon Tethys (right) has Odysseus Crater at 400 km (250 miles) in diameter, or about two-fifths of the moon's diameter.

and slide to the crater floor. After the first minute or so (after things stop falling) the crater is then subject to any other tectonic or climatic processes that can lead to alteration—uplift, erosion, flooding with water or lava, etc.

Although we see craters everywhere, large impact craters on small bodies really stand out. We often see evidence, on the opposite side (the antipodes of the crater), of stress fractures associated with the impact. These cracks suggest that impact shock waves traveled completely around the impacted body, sometimes many thousands of kilometers.

Many small bodies are fortunate even to remain intact at all! For example, the giant impact that produced the iconic Herschel Crater on Saturn's moon Mimas (also known as the Death Star moon after its ominous *Star Wars* look-alike) must have rung the small moon like a bell, nearly shattering it into pieces. One possible reason numerous other small bodies have also been spared similar fates is that some of the smallest bodies may actually be rather porous. Impact shock waves are more readily damped out in bodies that resemble Styrofoam more than solid rock. The impactor merely crushes material inward rather than blasting ejecta out.

However, the most extreme *individual* craters are so large that they often aren't immediately recognized as impact basins. The South Pole–Aitken Basin on our own Moon is one such structure. At 2,500 km (1,550 miles) in diameter and 12 km (7.5 miles) deep, it is the largest and perhaps oldest surviving impact feature on the Moon. Mars's Borealis Basin is truly gigantic—over 8,500 km (5,300 miles) in diameter! The status of Borealis as an impact basin remains in contention,

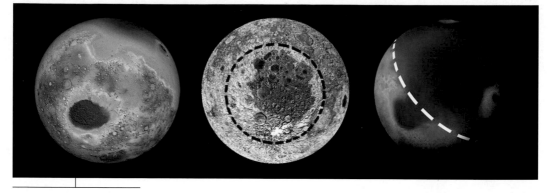

The three largest impact features in the Solar System: the 2,100-km (1,300-mile) Hellas Basin on Mars (left); the 2,500-km (1,550-mile) South Pole–Aitken Basin on our Moon (center); and the vast 8,500-km (5,300-mile) Borealis Basin of Mars (right).

but, if confirmed, it would be the largest impact feature in the Solar System *and* finally provide an explanation for the mysterious martian crustal dichotomy!*

But the grand prize for planetary welt marks must go to Jupiter's large moon Callisto, the most heavily cratered body in the Solar System. The cold, icy moon is so geologically inert that there has been very little in the way of tectonic modification. Callisto is the only relatively large body in the Solar System with no signs of extensive resurfacing—it has perhaps the oldest landscape in the Solar System. As a result, the moon's surface is very close to crater saturation—no more craters can be created without erasing other craters. Talk about leaving a mark or two ... or a few thousand!

Callisto, the most heavily cratered body in the Solar System.

*The northern hemisphere of Mars largely consists of relatively smooth lowlands some 2–5 km (1–3 miles) lower in elevation than older highlands in the southern hemisphere.

Part Two

Oceans & Water & Ice, Oh My!

Deepest Ocean—Europa

All these worlds are yours except Europa.
Attempt no landings there.
—Arthur C. Clarke, *2010: Odyssey Two*

In Arthur C. Clarke's 1982 novel *2010: Odyssey Two,* an alien intelligence broadcasts a warning against human interference with the primitive life-form developing in the ocean beneath Europa's icy surface. Although an alien message remains science fiction, nearly 30 years after Clarke wrote those words, NASA and ESA are planning a joint mission to the Jupiter system—and to Europa in particular. The primary science goals of the mission: characterize the surface ice, the subsurface ocean, and the moon's potential as an abode for life. Science fiction becomes science fact. How did Clarke know that we would find an ocean—much less search for life—on Europa?

Although he *was* a revolutionary thinker (he popularized the concept of geostationary communications satellites nearly 20 years before the first was built and launched), Clarke wasn't exactly imagining the future in this instance. Speculation about possible liquid oceans beneath the surfaces of the icy Jovian moons had begun a decade before in the early 1970s.

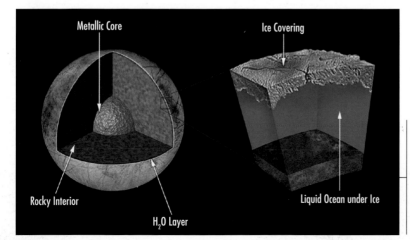

Metallic Core

Ice Covering

Rocky Interior

Liquid Ocean under Ice

H_2O Layer

Europa, Jupiter's sixth moon, has long been suspected of harboring an ocean beneath its icy crust. Measurements made by NASA's Galileo spacecraft suggest that Europa's subsurface ocean could be a layer of salty water 100 km (62 miles) thick that lies between the frigid surface ice and the rocky silicate mantle.

These ocean hypotheses were strengthened by more accurate measurements. Better determinations of mass and density were obtained during the Pioneer 10 and Pioneer 11 spacecraft flybys of Jupiter and its moons in 1973 and 1974. Meanwhile, ground-based spectroscopic measurements pointed to significant amounts of water ice in the Jupiter system. Theoretical models of the Galilean satellites predicted that while the moons' surfaces were indeed frozen solid, a significant amount of the water could still be found in liquid form beneath their icy shells.

Today, many think that subsurface oceans—layers of liquid water locked beneath frigid surface ice—may, in fact, be the most common form of ocean in the Solar System. There are indications that oceans may exist not only within some of the moons of Jupiter but also in those of Saturn, Uranus, and Neptune. Even dwarf planet Pluto and its moon Charon may harbor oceans beneath their frozen crusts. Only our own blue Earth has its liquid water exposed at the surface instead of trapped beneath ice.

The case for subterranean water is still mostly circumstantial, however. The best evidence yet for the existence of subsurface oceans comes from NASA's Galileo orbiter and its measurements of the magnetic fields around Jupiter's Galilean satellites.

As Europa, Callisto, and Ganymede (three of the four Galilean satellites) move through Jupiter's intense magnetic field, *induced* magnetic fields are generated within them. Induced fields can be produced only by electrically conducting material inside the moons. The most likely conductors for these moons? Layers of salty water beneath their outer ice shells. That's right, subsurface oceans!

For Callisto and Ganymede, the evidence is not completely conclusive, and interpretation of Ganymede's induced field is further complicated by that moon's own internally generated magnetic field. Any liquid water in these moons would have to be several hundred kilometers below the surface and trapped between layers of ice. The sizes of these potential ocean layers are not very well constrained. They could be anywhere from 10 to 200 km (6–120 miles) thick depending on the salts or other chemicals dissolved in the underground seas.

Detection of Europa's ocean, on the other hand, is much more definitive. Gravity measurements for flybys of the Galileo spacecraft

(Opposite page) Ice, oceans, and the potential for life will be the focus of the joint NASA/ESA mission under development to explore the Galilean satellites of Jupiter—especially Europa (foreground) and Ganymede (center right). Note the magnetic field lines emerging from Europa and Ganymede. Closer to Jupiter, Io interacts with the planet's magnetic field to produce intense auroras.

Models of the interiors of Europa (left), Callisto (center), and Ganymede (right) showing the approximate sizes and locations of the subsurface oceans (in blue) within each moon. Europa and Ganymede have differentiated into distinct iron cores and silicate mantles beneath icy crusts. Callisto, only partially differentiated, has an interior made up of a mix of rock and ice.

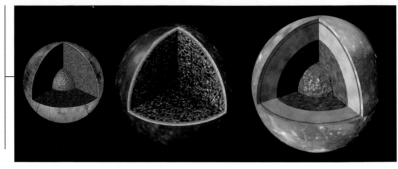

imply that Europa's interior is differentiated into a metallic iron core, a rocky silicate mantle, and an ice/water outer shell. A 4:2:1 orbital resonance with Io and Ganymede (Io completes four orbits, Europa two orbits, and Ganymede one orbit in the same period) ensures that Europa has a continuing supply of tidally generated heat. In addition, Europa's relatively large silicate mantle means that heat from radiogenic sources could be significant.

This heating within Europa's interior is more than enough to keep the moon's water from freezing solid. Estimates place the thickness of the outer icy crust at a few tens of kilometers and the

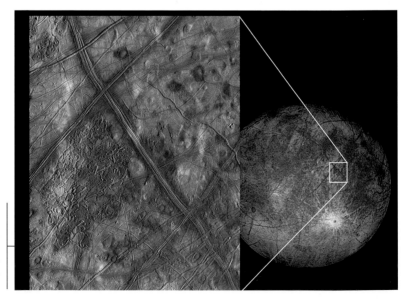

Geologic evidence supporting a subsurface ocean on Europa includes "ice rafts" — broken blocks of Europa's icy crust that appear to have floated into new positions. The reddish areas are non-ice material and appear to be associated with geologic activity.

Red ice at Blood Falls near Lake Bonney Antarctica. The red color comes from iron salts dredged up from a pocket of salty water trapped below Taylor Glacier. Extremophiles—organisms that live in extreme conditions—were recently discovered living in the isolated, salt-rich, oxygen-poor pools of brine 400 m (0.25 mile) beneath the ice. Could this be a model for life in Europa's salty subsurface seas?

salty, conductive ocean below it at around 100 km (62 miles) thick, making Europa's ocean the deepest (thickest) in the Solar System. By comparison, Earth's meager oceans are only 11 km (6.8 miles) at their deepest and average closer to 4 km (2.5 miles) in depth. In total volume, Europa's global subsurface ocean could hold roughly twice the amount of liquid water in all of Earth's oceans combined.

Although most planetary scientists agree that an ocean within Europa is a near certainty, debate continues over the thickness of the outermost icy shell. Fortunately, the NASA/ESA Europa Jupiter System Mission is designed to resolve such debate. Although it doesn't quite match Clarke's fictional time line, the mission is being planned for launch in 2020. Two separate spacecraft will fly by each of the Galilean satellites. ESA's component, the Jupiter Ganymede Orbiter (JGO), will eventually settle into orbit about Ganymede, while NASA's Jupiter Europa Orbiter (JEO) will end up in orbit about Europa.

The thickness of the icy crust may have significant implications for life. Europa's salty ocean offers a possible ideal habitat for extremophile organisms—protected from hazardous radiation by a layer of ice, warmed from within by tidal flexing, and in contact with the moon's rocky interior for a source of minerals. Europa's ocean is perhaps the most likely place to find life elsewhere in our Solar System. If the small icy moon were found to harbor life, that would be *deep* indeed.

Surfers travel around the world to catch the ultimate ride. But where should interplanetary surfers go to find the most righteous waves?

Surprisingly, choice surf spots in our Solar System are few. In the distant past, liquid water oceans likely covered the surface of Mars, but today Mars is a cold and dry place—lots of sand but no surf. And although Jupiter's moon Europa has a deep ocean capped by a thick layer of water ice, no liquid flows on the surface. Only two planetary bodies currently have liquid on the surface: Saturn's moon Titan and our own planet Earth.

Scientists had long hypothesized that Titan might have large oceans. Titan's surface temperature (a balmy –179°C or –290°F) is

Surf's up! Earth is the only planetary body with most of its surface covered in liquid.

just right to support liquid methane and ethane (water is frozen solid at this temperature). Remote measurements suggested an abundance of these hydrocarbons in the atmosphere, hinting at a substantial hydrocarbon cycle much like the water cycle on Earth. But to the dismay of surfers, recent observations from the Cassini-Huygens mission reveal that Titan is not an ocean world.

With liquid on its surface, Titan may be a prime surf location. This composite image depicts a lake on Titan (left) larger than Lake Superior on Earth (right).

There is still hope, however—numerous hydrocarbon lakes and rivers speckle the surface. Since some of Titan's lakes are as large as inland seas on Earth, it would come as no surprise to find, quite literally, Titanic waves pounding the giant moon's shorelines.

Yet for epic surf conditions, you need to look no farther than our home planet. Waves pummel the Earth's coastlines every day. Earthquakes beneath the ocean floor can produce tsunamis with wave heights larger than a 10-story building. These waves form

A massive tsunami obliterated the shoreline of Lituya Bay, Alaska, in 1958. The 520-m (1,700-ft) wave is the largest on record. Given the unpredictable nature of tsunamis, it is generally a bad idea to try to surf them.

abruptly and can be quite dangerous. In 2004, an estimated 230,000 people perished in a series of tsunamis produced by a 9.3-magnitude earthquake in the Indian Ocean, one of the worst natural disasters in history.

Ocean tides caused by the gravitational pull of the Moon and Sun produce a slow rise and fall of sea level usually every 12 hours. Although some radical surfers attempt to ride tidal bores as high tide flows upstream at the mouth of a river, most tidal waves occur too slowly to rip a major curl.

The most consistently rideable waves are produced by the wind. As air blows across the water, small ripples form and are pushed in the direction of the wind. These ripples combine with other ripples from different locations, sometimes canceling each other out and at other times adding together to form larger waves. Over time, these waves continue to grow and become organized in regular patterns called swells. Swells travel great distances across the ocean and can often take weeks to reach land.

To produce the largest waves, strong winds and a large fetch of open ocean are required. Luckily, those conditions occur readily on Earth. Consider, for example, the Pacific Ocean: Violent tropical typhoons and powerful midlatitude cyclones routinely sweep across Earth's largest body of water. With few obstacles in their way, these storms easily stir up the ocean surface into rough seas. Wind-generated waves have plenty of time to transition from turbulent choppy waves to relatively smooth (yet large) undulations as they propagate over the vast distances of the Pacific.

When a wave reaches shallower water, things become nasty again. Essentially, the wave "trips" on the ocean floor and begins to overturn as the top part of the wave overtakes the bottom part. A huge amount of energy—energy that originated in the open ocean hundreds or thousands of kilometers away—is dumped on the shore. This energy

World-record 21-m (70-ft) wave ridden by Peter Cabrinha in 2004 at the notorious Jaws surf spot on Maui. The waves are so monstrous that surfers must be towed in by Jet Skis. Ken Bradshaw purportedly rode a wave in Hawai'i over 25 m (83 ft) high in 1998, but there is no photographic evidence to confirm the height.

The authors on their surfboards at the El Porto surf break in Manhattan Beach, CA, patiently waiting for the "Big One." *Good* surfers don't even bother getting in the water with flat conditions like this.

incessantly erodes the coastline, year after year, millennium after millennium. And surfers use it to get the ultimate thrill.

The most famous surf breaks in the world take advantage of these conditions: Bonzai Pipeline (north shore of Oahu), Jaws (Maui), Cortes Banks (an underwater mountain range 160 km or 100 miles off the coast of southern California), Mavericks (northern California), and Teahupoo (Tahiti). On weak days, wave heights may only reach 2 m (6.6 ft). But on the hairiest days, waves can rise 10 times higher and catapult surfers with 100 times more power. Big-wave surfers drop down the vertical face of a wave at death-defying speeds of over 80 km/hr (50 mph). When they wipe out, it's like hitting a brick wall.

Surfing the local breaks still not extreme enough for you? Due to Titan's lower gravity, wind-driven waves could be up to 10 times larger than those on Earth under the same wind conditions. So who knows? We may yet discover that Titan experiences the gnarliest surf in the Solar System.

Extreme Facts about Earth's Ocean

- 71% of the Earth's surface is water.
- The Pacific Ocean itself covers one-third of the Earth's surface, more than all of the continents combined.
- The top 3 m (10 ft) of the ocean store more energy than the entire atmosphere (including lightning storms, tornadoes, and hurricanes!).
- The average depth of the ocean is 3,790 m (2.3 miles), and the deepest point is the Mariana Trench in the western Pacific at 11,034 m (6.7 miles).
- 80% of all life on Earth is found under the ocean surface.
- Less than 10% of the ocean has been explored by humans.

If you wanted to make a Solar-System-sized snow cone (and who wouldn't?), you would definitely need a lot of ice. Fortunately, our cosmic neighborhood is teeming with the stuff. Ice can be found almost everywhere . . . well, maybe not on the stifling-hot greenhouse that is Venus. But it *is* at the poles of Mercury, Earth, the Moon, and Mars. It whizzes through the Solar System on comets. Ice is a large component of the satellites of the outer planets, the Kuiper Belt, and the Oort Cloud. If there is one thing our Solar System *doesn't* lack, it's ice!

But if you were making an immense snow cone you wouldn't want just any old ice. You would want large quantities of relatively pure water ice. Something not contaminated with rocks, dust, or metals. No ammonia or frozen methane. You would need the biggest ice cubes in the Solar System.

You might be tempted to just hang around Earth and catch comets as they swing by, but cometary ice is probably not the kind of ice you would want to use for your dessert. Comets are called dirty snowballs for a reason: you will find quite a bit of stuff other than ice in them.

No, you would much rather have ice that is considerably cleaner. The polar caps of the terrestrial planets—especially Earth's—seem appealing at first. However, the rocky inner planets don't have nearly enough ice. You need more!

We should head a little farther past the frost line, which lies roughly between the present-day orbits of Mars and Jupiter. Inside the frost line (nearer the Sun) it is too warm for ices to form easily, so rocks and metals dominate—the inner planets are all rocky bodies with iron cores. Outside the frost line, various volatiles (mostly water) can readily form ices.

(Opposite page) Selected moons of the outer Solar System shown along with Earth for scale. While most of these satellites have significant portions of water ice (Io is a notable exception), three moons of Saturn (Mimas, Tethys, and Iapetus) have densities close to that of solid water, making them the Solar System's biggest ice cubes.

Since you need more ice, let's think big. Ganymede, perhaps? Ganymede, one of the four Galilean satellites of Jupiter, is the largest natural satellite in the Solar System. The satellite's *surface* could be as much as 90% water ice. The rest of Ganymede, however, isn't quite so icy; it's probably only 40% ice overall. Plus, based on the intrinsic magnetic field observed by the Galileo spacecraft, the large moon

Selected Moons of the Outer Planets

Jupiter — Io, Europa, Ganymede, Callisto

Saturn — Mimas, Enceladus, Tethys, Dione, Rhea, Titan, Hyperion, Iapetus, Phoebe

Uranus — Puck, Miranda, Ariel, Umbriel, Titania, Oberon

Neptune — Proteus, Triton, Neried

Earth (for scale)

Although Jupiter's moon Ganymede has a lot of ice, over half of the largest natural satellite in the Solar System is rock.

likely has an iron core in addition to its silicate mantle. With such a large crunchy center, you would need to take extra care not to push the moon's rocky parts and iron parts through your gigantic ice shaver.

Maybe Ganymede is too big, then. In fact, most of the large icy satellites have a rocky/iron center. Let's look, instead, for smaller icy moons that are almost *all* ice.

How can we tell which bodies are mostly ice? By comparing the average density of the icy satellite to the densities of water (1 g/cm^3) and iron-rich silicate rocks (3–3.4 g/cm^3), we can get a general idea of which moons are mostly rocky and which are mostly icy.

Pure water ice—at around 0.92 g/cm^3—is slightly less dense than water (that's why ice cubes float in your drink); but we can consider densities a little larger. Regular water ice can be transformed into slightly more dense versions under colder temperatures and higher pressures—there are 15 different known forms of ice (the ice in your kitchen freezer is ice I). We don't, however, want the density to be too much smaller than 1 g/cm^3, since that would imply lots of empty space inside the moon. You don't want to make your snow cone from a crumbly slush pile; you want to be able to shave ice off a nice solid chunk.

There are at least 164 known moons orbiting the giant outer planets. If we consider only those icy moons with a density near that of water, we're left with a much shorter list. Of Jupiter's 63 moons, only Amalthea is a likely candidate (density 0.849 g/cm^3), but the small, irregularly shaped moon's reddish tinge means that there's something on the surface that might not be so tasty for our snow cone. The color could be due to sulfur compounds originating from stinky, volcanic Io. Rotten-egg-flavored snow cone, anyone? We didn't think so.

While distant Neptune has no satellites with densities near that of water ice (that we know of thus far), the other ice giant, Uranus, has one possible candidate for a pure ice moon: tiny, fractured Miranda.

Although composed mostly of water ice, Saturn's moon Iapetus has a dark side. The side of the moon pointed in the leading direction of its orbit is covered with dark carbonaceous material, an extra flavoring for your snow cone.

Since Miranda isn't large enough to form the more dense, higher pressure phases of ice, its slightly higher density of 1.214 g/cm³ means that the Uranian moon probably has at least 20% rock inside.

Ringed giant Saturn is a much more fertile hunting ground for giant ice cubes. First, there are those rings. Spectroscopic measurements indicate that the ring material is nearly 99.9% pure water ice— that's good. However, as big and as brilliant as they appear, Saturn's lovely rings contain only a little more ice than Earth's polar caps and ice sheets. That's still not enough ice for the snow cone we want!

Luckily, at least three of Saturn's moons have densities very close to that of water ice: Iapetus (1.083 g/cm³), Mimas (1.150 g/cm³), and Tethys (0.973 g/cm³). As the third largest of Saturn's moons, Iapetus contains a lot of ice; but here, too, more precise estimates place the rock fraction closer to 20%. Tiny Mimas, on the other hand, doesn't

The features of Saturn's fifth largest moon, Tethys, are given names from Homer's *Odyssey*. In this mosaic of the southern hemisphere as seen by the Cassini spacecraft, the massive crevice Ithaca Chasma stretches up across the disk of the moon. This moon may very well be the largest ice "cube" in the Solar System.

provide nearly enough ice—it contains less mass than the rings of Saturn.

If you're after the most ice with the least rock—and you are!—then Tethys is the way to go. With only 3% rock, Tethys may, in fact, be the largest ice cube (and here we mean "cube" in the spherical sense, of course) in the Solar System. This icy satellite contains more than 20 times the amount of ice on Earth. The Tethyan surface is dominated by two major features: Ithaca Chasma—a vast crack in the ice, possibly formed during the solidification of a subsurface ocean—and Odysseus Crater—a massive impact basin. As an added bonus, the giant ice cube comes with two smaller Trojan moons (possibly also made entirely of water ice), Telesto and Calypso, which co-orbit Saturn 60° ahead of and behind Tethys.

Enjoy your giant snow cone from Saturn! Now where can we find enough flavored syrup to squirt on top?

Jekyll and Hyde of the Solar System— Dirty, Icy Comets

The Solar System gods must laugh at our relentless attempts to classify objects and fit them into neat little boxes. Just when we think we understand something, new discoveries are made and the boxes aren't so neat anymore. We then build stronger boxes to accommodate the new information, yet even these boxes will eventually be replaced by more robust theories. Ah, the progress of science . . .

Take the traditional view of asteroids and comets. For years, asteroids have been characterized as inert chunks of rock inhabiting the inner Solar System and comets as icy bodies from the outer Solar System. Asteroids were thought to be relatively warm but geologically dead, whereas comets were thought to be cold but active.

Comets have impressive comas and/or tails, but asteroids do not. The coma is a tenuous atmosphere of volatiles like water and carbon dioxide spewed out from the solid cometary nucleus. There are two

Comet McNaught (officially called C/2006 P1) observed from Paranal, Chile, on January 20, 2007. The expansive curved dust tail follows the orbit of the comet. It contains unusual striations, perhaps caused by uneven fragmentation of the cometary nucleus. A fainter gas tail to the left of the dust tail points away from the Sun. A crescent moon (overexposed in this image) hangs near the horizon. Comet McNaught is a nonperiodic comet, never to return to the inner Solar System. It was the brightest comet in the night sky since 1965.

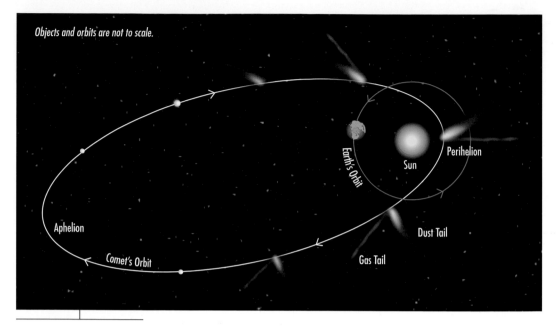

Objects and orbits are not to scale.

Earth's Orbit

Sun

Perihelion

Aphelion

Comet's Orbit

Gas Tail

Dust Tail

A periodic comet has a highly elliptical orbit that places it nearest the Sun at perihelion and farthest away from the Sun at aphelion. Near perihelion, the comet moves quickly in its orbit and brilliant cometary tails are produced as the comet heats up. Near aphelion, the comet moves more slowly and has no tails.

types of tails: the gas tail, a long extension of the coma swept away by the solar wind, and the dust tail, a trail of debris left in the orbital path of the comet. With gas and dust tails made of different compounds and therefore different colors, comets create some of the most beautiful sights in the night sky.

But things are a bit fuzzier (and if you look at a comet through a telescope, it *does* look fuzzy) than this simple view. In the early 1950s, eminent scientist Fred Whipple suggested that comets were not pure ice but instead a mixture of ice and rock—the "dirty snowball" hypothesis. More recent discoveries show that comets are dirtier and asteroids more icy than previously suspected: ESA's Giotto spacecraft flew by Comet 1P/Halley (yes, *that* Comet Halley!) in 1986 and detected a cometary nucleus as dark as coal; NASA's Deep Space 1 spacecraft discovered the darkest surface in the Solar System on Comet 19P/Borrelly; a new type of small body (a main-belt comet or outgassing asteroid, take your pick) was discovered in the main asteroid belt between Mars and Jupiter in 2004; and NASA's Deep Impact probe slammed into Comet 9P/Tempel in 2005, blasting more dust and less water into space than anticipated.

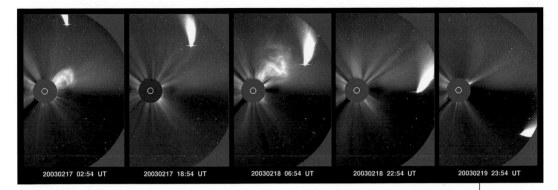

| 20030217 02:54 UT | 20030217 18:54 UT | 20030218 06:54 UT | 20030218 22:54 UT | 20030219 23:54 UT |

In other words, comets may be much more like asteroids—"icy dirtballs" rather than "dirty snowballs"—and the traditional line between comets and asteroids has become blurry.

The most extreme comets confuse matters even further. Sometimes behaving like comets, sometimes like asteroids, these small bodies are the Dr. Jekyll and Mr. Hyde of the Solar System.

Unlike the main planets with nearly circular orbits, comets travel through the Solar System in highly eccentric orbits around the Sun. Some comets have such extreme trajectories—parabolic or hyperbolic orbits—that they will pass close to the Sun only once before being ejected from the Solar System altogether. Both periodic and nonperiodic comets have cold origins in the far reaches of the Solar System: the Kuiper Belt and the Oort Cloud. These distant bodies become perturbed by something (perhaps by a passing star or by our Solar System grazing through the Milky Way's dusty spiral arms) and their orbits become more eccentric.

A good example is Comet 81P/Wild. For most of its long life, it behaved like the staid, mild-mannered Dr. Jekyll of comets—cold, slow, and quiet, more like an icy asteroid than a comet. Comet Wild led a calm existence in the outer Solar System, with relatively slow orbital motion and little to no outgassing. In fact, it was so benign that it remained undiscovered. Then in 1974, the small body passed too close to Jupiter. The giant planet's massive pull careened the

Observations of Comet NEAT (C/2002 V1) over three days by the Solar and Heliospheric Observatory (SOHO) as the comet passes near the Sun. The Sun is obscured in these images. Note large coronal mass ejections from the Sun in the first and third images.

Temperature map of Comet 9P/Tempel from the Deep Impact spacecraft: the comet is sizzling hot in the sun and freezing cold in the shade (sunlight is coming from the right in this image). Although farther away from the Sun, the day side temperature of Comet Tempel equals the highest surface temperature ever recorded on Earth.

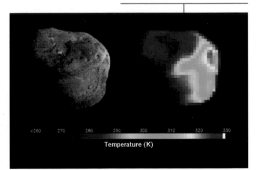

| <260 | 270 | 280 | 290 | 300 | 310 | 320 | 330 |
Temperature (K)

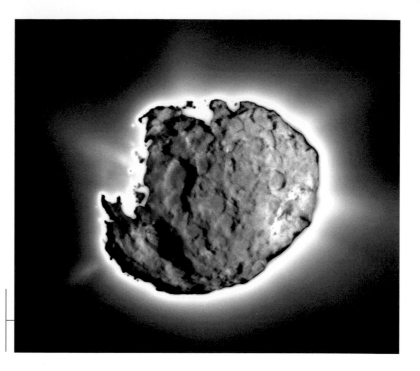

Hello wild Mr. Hyde: this composite image of Comet 81P/Wild shows both the solid nucleus of the comet and bright jets of outgassed material that produces the coma.

small comet into a new orbit (and into a new personality), sending it toward the inner Solar System.*

As the comet approached the Sun, the leading side became warmer and warmer. Frozen ice began sublimating into gas, and strong jets of warm gaseous volatiles erupted from the nucleus. A thin coma formed around the nucleus, becoming slightly elongated into a gas tail by the solar wind. Increased activity on the surface blew rocky particles off the nucleus, leaving a trail of dust in the comet's wake. As the comet got closer to the Sun, it experienced temperatures higher than any it had felt in nearly 4.5 billion years. Comet Wild zoomed through the Solar System faster than it ever had before.

The transformation into the Mr. Hyde of comets—hot, fast, and dirty—was complete. After years of obscurity, Comet Wild was finally

---

*Comet 81P/Wild's new egg-shaped orbit has an eccentricity of 0.538 with perihelion inside Mars's orbit and aphelion outside Jupiter's orbit. By comparison, Earth's nearly circular orbit has an eccentricity of 0.017. Nonperiodic comets have eccentricities of 1.0 or greater.

discovered in 1978 due to its new, er, "wild" lifestyle (okay, okay, the name is actually pronounced *vilt* after its Swiss discoverer).

Then, after its closest approach to the Sun, the comet began transforming back into Dr. Jekyll. Colder temperatures shut down sublimation, the coma and tails disappeared, and Comet Wild behaved again like an icy asteroid. This particular comet has undergone the personality switch just a few times, as it has completed only five orbits since 1974.

Some comets have a longer history of personality disorder—Halley's Comet has passed close to the Sun over 100 times in its 86-year elliptical orbit. Other comets get only one chance to be hot and fast—Comet McNaught's parabolic orbit allows just one close approach to the Sun.

And then there is Chiron. A member of a group of small Solar System bodies called Centaurs with orbits between Jupiter and Neptune, Chiron was initially classified as an asteroid upon its discovery in 1977. Then in the late 1980s, Chiron brightened significantly and developed a cometary coma. Chiron is now officially both an asteroid (2060 Chiron) and a comet (Comet 95P/Chiron). And Chiron is not alone—other Centaur "asteroids" have recently exhibited outgassing and cometary characteristics.

Our neat boxes for small Solar System bodies are being torn apart. In reality, there is likely a broad spectrum of small Solar System bodies, from rocky objects to icy bodies and everything in between. They may be sometimes calm and quiet and at other times angry with jets and geysers. Some "asteroids" may be old comets with no volatiles left, while other "asteroids" may be comets waiting to happen.

We don't know what the new boxes will look like. We may find small bodies with multiple personalities even more peculiar than our current Jekyll and Hyde comets. Like the famous doctor and his alter ego, our understanding of the Solar System is undergoing an extreme transformation.

The Sky Is Falling!—Dry Ice Caps of Mars

CO$_2$ dry ice at the north pole of Mars really stirs things up: three swirling dust storms can be seen here along the edge of the seasonal ice cap in early northern spring. The temperature difference between the ice cap and the adjacent darker ground causes strong winds to flow off the ice cap. This mosaic of images was taken in May 2002 by Mars Global Surveyor's (MGS) Mars Orbiter Camera (MOC).

What does the haunted house attraction at an amusement park have in common with the planet Mars? Sure, they are both cold and dusty. And yes, they may both harbor ghosts of past life. But there is one feature that lends atmosphere to both places and makes them quite unusual: dry ice.

Dry ice is frozen carbon dioxide. On the surfaces of Earth and Mars, carbon dioxide is able to exist as a gas or a solid (dry ice) but not as a liquid. Liquid CO$_2$ can occur only at pressures higher than 5.1 atmospheres (1 atmosphere is the average atmospheric pressure at sea level on Earth). In fact, when dry ice "melts," it skips the liquid phase

entirely and turns directly into vapor in a process called sublimation. Unlike H_2O, a dipole molecule with positively and negatively charged ends, CO_2 is electrically neutral and forms only weak electrical bonds between molecules. It takes very little thermal energy to break these molecular bonds, so solid carbon dioxide easily turns into carbon dioxide gas.

In a haunted house, dry ice is used to create an eerie mist, perhaps bubbling from a witch's cauldron or rising out of a distant graveyard. This mist forms because sublimation of dry ice releases extremely cold molecules of CO_2 vapor—the surface temperature of dry ice is –78.5°C (–109°F)—which cool the nearby air. Ambient water vapor rapidly condenses in this superchilled air to form an ethereal fog.

This same process happens on Mars, except that it involves the planet's polar ice caps and the entire martian atmosphere. Temperatures at the poles of Mars are, well, ice cold—warm summertime temperatures reach only a frigid –95°C (–139°F)—leading to residual (permanent) ice caps at each pole.

The north residual polar cap, composed entirely of water ice roughly 3 km (1.8 miles) thick, covers an area about the size of the U.S. state of Texas. The ice rests in a deep basin that could have been created by an ancient meteor impact or internal tectonics. Sand dunes over 500 m (1,650 ft) high surround the permanent ice field. These towering dunes are continually shaped by strong winds flowing off the ice cap, similar to the cold downslope katabatic winds of

The northern (left) and southern (right) residual polar ice caps on Mars at minimum size during their respective summers. The northern permanent cap is 1,100 km (680 miles) across and consists of water ice. The smaller southern permanent cap contains both water and carbon dioxide ice and is 420 km (260 miles) across. During winter, the entire area in each image would be covered in CO_2 frost, creating a much larger seasonal ice cap of dry ice.

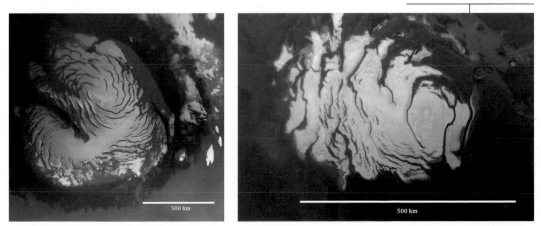

500 km

500 km

This Hubble Space Telescope image shows polar-hood water ice clouds surrounding the north polar cap and the large seasonal carbon dioxide ice cap at the south pole.

The surface of the southern residual ice cap resembles Swiss cheese. The "holes" are caused by sublimation of dry ice, revealing water ice below. This Mars Global Surveyor MOC image spans roughly 900 m (3,000 ft) across.

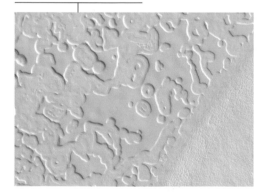

Antarctica. The martian winds also may help carve landmarks unlike any glacial feature found on Earth—deep spiral chasms in the ice cap itself.

The southern permanent ice cap is quite different from that of the north. Since the south pole is higher in elevation by about 6 km (3.7 miles), temperatures there are cold enough for carbon dioxide ice to persist throughout the year. Although the southern ice cap was once thought to consist entirely of CO_2, recent radar measurements by ESA's Mars Express spacecraft show a thick water ice layer—over 3 km (1.8 miles) deep—with a thin veneer of solid carbon dioxide only 8 m (25 ft) thick. Interestingly, the center of

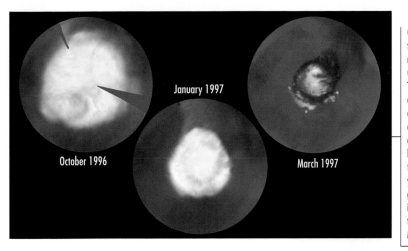

October 1996

January 1997

March 1997

Changes in the amount of dry ice at the north pole in early spring (left), midspring (center), and early summer (right) detected by the Hubble Space Telescope. Note that seasons on Mars do not coincide with our own here on Earth. Such variations in polar dry ice drive seasonal fluctuations in atmospheric carbon dioxide, shown here from Viking Lander measurements from 1976 to 1980 (two martian years). Atmospheric pressure is greater—more ice cap sublimation—in southern summer (northern winter) than in northern summer because of Mars's eccentric orbit.

this ice cap is offset from the geographic pole by about 150 km (93 miles), a result of a snowy low-pressure weather system bizarrely trapped in one location by two adjacent impact basins.

But the really cool (if −95°C can be considered merely cool) stuff happens as Mars changes seasons. As summer moves into fall, ghostly clouds of water and carbon dioxide ice surround the permanent ice cap in a polar hood. CO_2 snowfall blankets the permanent ice cap. As winter settles in, CO_2 frost extends from the pole to 60° latitude. This large *seasonal* dry ice cap is over 1 m (3.3 ft) thick, with deeper snowdrifts on mountain slopes and gullies.

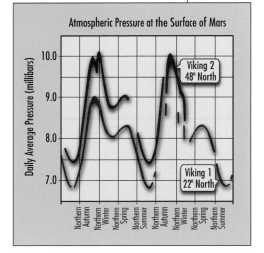

Atmospheric Pressure at the Surface of Mars

Daily Average Pressure (millibars)

Viking 2
48° North

Viking 1
22° North

Northern Autumn | Northern Winter | Northern Spring | Northern Summer | Northern Autumn | Northern Winter | Northern Spring | Northern Summer

Where do all this CO_2 ice and snow come from? Unlike Earth's atmosphere, in which carbon dioxide is only a minor constituent, the Martian atmosphere is over 95% CO_2. As sunlight decreases at the pole in autumn and temperatures plummet, carbon dioxide gas begins to freeze. On Mars, the sky is indeed falling—roughly 25% of the atmosphere precipitates out as snow onto the winter seasonal ice cap!

When sunlight returns to the pole and winter turns to spring, the seasonal dry ice cap starts sublimating and carbon dioxide is returned to the atmosphere. By the middle of summer, only the residual ice cap remains. Because the orbit of Mars places the Red Planet closer to the Sun in southern summer, more CO_2 ice sublimates in southern summer than in northern summer. The southern residual ice cap is therefore smaller, and atmospheric pressure reaches a maximum during this time.

Mars may not be the only planetary body in the Solar System where the sky is falling. The very thin atmosphere of Pluto may also freeze out onto the surface as the dwarf planet moves farther away from the Sun in its highly elliptical orbit. Similar processes are likely occurring on other bodies in the outer Solar System as well. But in the inner Solar System, Mars stands alone in its sublime (okay, sublimating) ways—the largest pack of dry ice in the Solar System, mystical clouds that envelop the poles, and an unrelenting falling sky that would make the infamous Chicken Little tremble.

Part Three

Wild, Wild Weather

Longest-Lived Storm—Jupiter's Great Red Spot

The year was 1665. Italian astronomer Giovanni Cassini turned his telescope to the planet Jupiter and made an amazing discovery—a large "permanent" spot in the southern hemisphere of the giant planet. Cassini and his successors observed the permanent spot regularly until 1712. Only sporadic observations occurred for the

True-color image of turbulent Jupiter in 2000 by the Cassini spacecraft, named after the discoverer of the enduring Great Red Spot.

Jupiter's Great Red Spot and a white oval as seen by Voyager 1 in 1979.

next 165 years, but the Great Red Spot has been systematically observed since 1878. Despite the large gap in the record, many astronomers think that the Great Red Spot has existed for more than 340 years, longer than the United States has been a country. Nevertheless, the Great Red Spot's definitive existence for the past 130 years makes it the Solar System's longest-lived storm.

Not only is the Great Red Spot long-lived, it is also immense and intense. Approximately three Earths would fit inside Jupiter's massive storm system. The Great Red Spot towers 8 km (5 miles) over the surrounding cloud tops, nearly the height of Mount Everest. The giant whirlpool is filled with thick, deep clouds and violent turbulence. Like hurricanes on Earth, the Great Red Spot possesses some of the fiercest winds on its home planet, with speeds in excess of 190 m/s (400 mph). Yet unlike Earth's low-pressure cyclones, Jupiter's high-pressure vortex spirals in the opposite direction (counterclockwise in the southern hemisphere).

For a moment, just imagine what such a storm would be like on Earth. Hurricane Katrina ravaged the Gulf of Mexico with maximum winds of 83 m/s (175 mph) and a storm surge of water over 8 m (27 ft)

Tower of power: dark areas indicate low clouds, pink indicates higher clouds, and white indicates thick, deep clouds in this 1996 false-color image of the Great Red Spot by the Galileo spacecraft.

Whew, just missed! Red Spot Jr. barely survived a close encounter with the Great Red Spot. The red spots are white in this infrared image taken on July 14, 2006, from the Gemini Observatory on Mauna Kea, Hawai'i.

high. The Great Red Spot is five times more energetic than Katrina[*] and would potentially produce a storm surge as tall as a 14-story building (55 m/180 ft high). Whereas Katrina lasted for only eight days, the Great Red Spot churns for years and years. The devastation by such a storm would be merciless.

So what makes this massive storm so long-lived? The Great Red Spot must have a continual source of energy to survive, and this energy comes primarily from two places: Jupiter's deep interior and nearby smaller vortices. Remarkably, Jupiter's interior supplies 70% more energy to the cloud tops than the planet receives from the Sun. Like a giant air compressor, gravitational contraction generates intense pressures and heat deep inside the planet. Powerful thunderstorms in Jupiter's atmosphere channel much of this heat to the

---

[*]Since kinetic energy depends on the square of the velocity, the Great Red Spot with speeds 2.3 times faster than Katrina produces energies over five times larger.

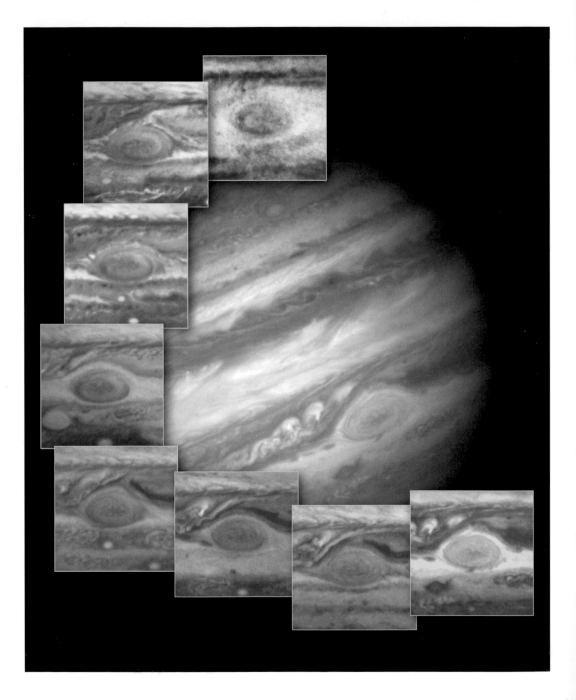

cloud tops. A single Jovian thunderstorm releases enough energy to power a typical U.S. household for over 75,000 years.

Because of this large amount of energy, Jupiter is a turbulent place. Whorls and eddies abound in the clouds, from relatively small swirling clouds to larger white ovals. But with a mammoth red vortex towering high above other clouds, rapidly spinning as it approaches smaller vortices, the smaller features don't stand a chance. The Great Red Spot is a voracious eater of smaller eddies. Red Spot Jr., a "small" red spot about half the size of the Great Red Spot, developed from the merger of three white ovals in the late 1990s (it later intensified and turned from white to red for unknown reasons). In 2006, Junior narrowly missed being devoured by its larger cousin. Although Junior escaped for now, the frequent consumption of energetic smaller vortices by the Great Red Spot has probably allowed the giant storm to exist for centuries.

Despite these mergers, the Great Red Spot has been shrinking. Although still enormous, it is only half as large as when observed in the late 19th century. Meanwhile, the speed of its circulating winds has increased dramatically. These variations are quite puzzling. Are they natural fluctuations that have occurred many times in the past, with the Great Red Spot repeatedly shrinking and swelling? Or will Red Spot Jr. continue to intensify and surpass its larger relative as Jupiter's "permanent" spot? Only time will tell what lies in store for the longest-lived storm in the Solar System.

(Opposite page) Hubble Space Telescope images of the Great Red Spot from 1992 to 1999 (counterclockwise from the top). Changes in color and turbulent flow suggest that the giant vortex may have vacillated often in strength during its long existence, possibly explaining the 165-year gap in observations. A light salmon-colored spot may have been too faint to detect with primitive telescopes.

Hurricane Wilma hits Cozumel on October 21, 2005. The central eye is roughly 70 km (43 miles) in diameter. Two days earlier, Wilma's eye had been only 5 km (3 miles) wide and had the lowest central pressure ever recorded for an Atlantic hurricane (882 mb).

A bank of clouds covers the sky. Winds steadily pick up speed throughout the day, changing direction ever so slightly. The ocean pounds the beach with irregular, vicious waves. Although not high tide, the ocean level is higher than normal, eroding old sandbars and creating new banks. Rain begins to fall, lightly at first, and then in sheets. Winds whip the pelting rain into stinging bullets. Debris flies through the air—metal panels, roof shingles, tree limbs. There is a brief reprieve in rain, but the wind accelerates and the ocean surges landward. The

next band of rain hits even harder, and the next band harder still. The deafening noise is unbearable . . . but it continues for hours.

Florida landfall of a 1945 category 4 hurricane.

And then it all stops. The wind and rain cease, and blue sky peeks out from behind the clouds. A weird sense of serenity envelops the air. This is the eye—the calm within the storm.

A short time later, the storm returns with full force. Strong winds and torrential rains again lash the surface. Water floods the streets. An isolated tornado may rip through the area. After a few hours, the onslaught of wind and water eventually dies down. The ferocious storm has passed.

Called hurricanes in the Atlantic and eastern Pacific, typhoons in the western Pacific, and tropical cyclones in the south Pacific and Indian Ocean, these storms are among the most extreme natural events on Earth. Each year, about 90 tropical storms form around the globe. Over half of these storms reach hurricane status (category 1+). Roughly 20 become major hurricanes (category 3+). The strongest tropical storm on record is Super Typhoon Tip (1979) with sustained winds of 85 m/s (190 mph) and a central pressure of 870 millibar (mb), the lowest surface pressure ever recorded.

Hurricanes rotate in a cyclonic direction, counterclockwise in the northern hemisphere and clockwise in the southern hemisphere. Within the circulating winds, strong updrafts form multiple bands of rain. The most intense area is the innermost rainband, or eyewall. As horizontal winds reach their maximum in the eyewall, atmospheric pressure drops within the eye itself. The warm, moist air in the eyewall is exhausted outward at the top of the hurricane, but some air returns to the surface through the center. This descending air inhibits cloud formation and clears the sky, producing

Eyewall of Hurricane Katrina from a NOAA WP-3D Orion hurricane hunter aircraft on August 28, 2005.

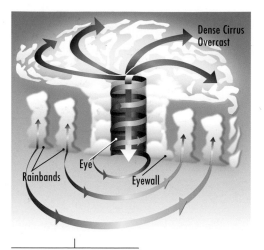

Cross section of typical hurricane structure. The fiercest (ill) winds circulate in the stormy eyewall, while the calmest (still) winds slacken nearby in the clear eye.

calm and deceptively benign conditions within the eye.

Perhaps the most powerful heat engines on Earth, hurricanes transport a lot of energy through the atmosphere. For hurricanes to form, sea surface temperatures must be at least 27°C (80°F) over a depth of 50 m (160 ft). This warm pool of water promotes evaporation. When water vapor condenses into clouds, energy originally absorbed during the evaporation of water at the surface is released into the atmosphere. This heat is often concentrated in "hot towers"—tall, water-heavy thunderclouds that penetrate into the stratosphere 15 km (9 miles) above the surface. Amazingly, the amount of heat transferred by a single hurricane in one day equals the caloric content of over 15 trillion slices of apple pie (it takes a lot to sustain a hurricane through a long day, after all).

Hurricanes inflict considerable damage to ecosystems, buildings, and human life. During Hurricane Katrina (2005), winds over 58 m/s (130 mph) persisted for many hours, rain rates exceeded 5 cm (2 in) per hour, and sea level rose by as much as 8 m (27 ft). Coastal ecosystems, forests, and freshwater bodies were significantly modified or destroyed. In 2005, estimated damages in the United States

Saffir-Simpson Hurricane Intensity Scale				
Category	Wind Speed (m/s)	Pressure (mb)	Storm Surge (m)	Description
Depression	< 17	—	—	—
Tropical Storm	17–32	—	—	—
1	33–42	> 980	1.5	Minimal
2	43–49	965–979	2.0–2.5	Moderate
3	50–58	945–964	2.5–4.0	Extensive
4	59–69	920–944	4.0–5.5	Extreme
5	> 70	< 920	> 5.5	Catastrophic

from hurricanes exceeded $120 billion and nearly 2,300 people died. The deadliest known tropical storm occurred in Bangladesh in 1970 when a strong storm surge flooded the low-lying country in the early morning. Over half a million people perished, many while asleep.

With the advent of satellite imagery, hurricane hunter aircraft, and sophisticated computer models, scientists have learned much about hurricane formation. Forecasts of intensity and hurricane track have improved dramatically over the past two decades, but many mysteries remain. We will have plenty of opportunities to learn. With global warming, tropical storms will likely become more frequent and intense in the future. The most ferociously ill yet calmly still storm may be yet to come.

Hurricane Katrina, a category 5 storm, had maximum sustained wind speeds of 78 m/s (175 mph) on August 28, 2005. The Tropical Rainfall Measuring Mission (TRMM) satellite looks beneath Katrina's clouds to estimate rainfall amounts. Blue indicates at least 0.64 cm (0.25 in) of rain per hour, green at least 1.3 cm (0.5 in) per hour, yellow at least 2.5 cm (1.0 in) per hour, and red over 5.1 cm (2.0 in) per hour. Katrina made landfall on the Louisiana coast nearly 14 hours later with a storm surge in Mississippi of 8 m (27 ft).

Neptune—a giant gaseous planet whose warm interior is . . . *ice*. A planet so distant that the Voyager 2 spacecraft, traveling at over 42,000 km per hour (26,000 mph), took 12 years to get there. The amount of sunlight arriving at Neptune is so paltry that high noon resembles twilight on Earth. And yet to the surprise of planetary scientists, this giant planet exhibits the fiercest planetary winds ever observed.

At first glance, Neptune's appearance is benign—deceptively so. The deep methane-blue planet exudes a certain serenity and calm, but closer inspection shows a much more violent world than initial impressions suggest: Neptune sustains winds near the equator of

The Voyager 2 spacecraft captured this image of a cold planet with wildly wicked winds as it approached Neptune in 1989.

450 m/s (1,000 mph)! These winds are somewhat similar to trade winds in the tropics on Earth but over 60 times faster. Storms with white icy methane clouds and dark eddies come and go in these winds like ghosts on a speeding train.

One of the most amazing discoveries by the Voyager 2 spacecraft in 1989 was the presence of a hurricane-like storm in the southern hemisphere of Neptune that has been dubbed the Great Dark Spot. Similar in structure to Jupiter's Great Red Spot, the Great Dark Spot could engulf the entire Earth. Wind gusts in excess of 600 m/s (1,340 mph)—the fastest winds ever measured on any planet—were detected near this turbulent vortex.

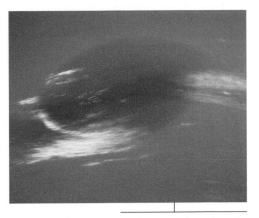

Neptune's Great Dark Spot observed by Voyager 2 in 1989. The deep blue color indicates methane absorption, and the white clouds are methane ice. Hubble Space Telescope images in 1994 indicated that the Great Dark Spot had vanished.

It is no surprise that the giant planets (Jupiter, Saturn, Uranus, and Neptune) experience large winds—after all, their lack of a solid surface reduces friction on the atmosphere. Yet not all giant planets behave alike. The atmospheres of Jupiter and Saturn have multiple zonal (east-west) jets—which we see as alternating bands of color. These high-speed jets alternate directions and exhibit strong turbulent eddies at their boundaries.

Just how fast are these jets? To get a good answer, you have to know the rotation rate of a planet's interior and then calculate how fast the atmosphere is moving relative to it. Since the gas giants do not have a solid surface, the interior rotation is usually estimated from magnetic field and radio wave measurements. Jupiter's jets are a relatively sluggish 100 m/s (220 mph), but Saturn's equatorial winds have been clocked as high as 470 m/s (1,050 mph) based on rotation measurements from Voyager spacecraft in the early 1980s. More recent estimates in 2009 of Saturn's rotation (Saturnian days are 5 minutes shorter than previously believed) downgrade the wind speeds to a mere 395 m/s (884 mph), not quite fast enough to overtake Neptune's speedy gusts.

In contrast, Uranus and Neptune lack multiple jets with sharp boundaries. Instead, each of these more distant giant planets shows a westward equatorial jet that transitions into a single eastward jet

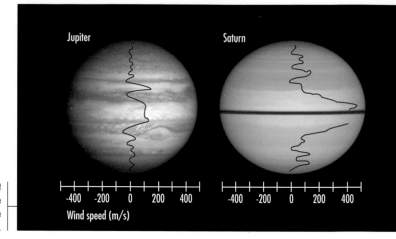

Zonal winds (jets) on the four giant planets. Saturn and Neptune have the strongest jets, but Neptune's winds are even faster near the Great Dark Spot.

toward the pole in each hemisphere. This disparate behavior is likely caused by two important factors: internal heating and rapid rotation.

Internal heating significantly impacts atmospheric circulations on every giant planet except Uranus. Because of their large distances from the Sun, the amount of sunlight reaching the giant planets is quite small—not nearly enough to drive extreme weather and high winds. The planets get an extra boost of energy from their deep interiors, either from gravitational contraction from the planet's formation (Jupiter) or from heat released by differentiation (raining out

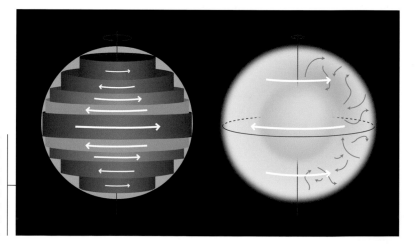

Rapidly rotating planets like Jupiter and Saturn (left) have alternating rotating cylinders to produce jets, while more slowly rotating planets like Uranus and Neptune (right) have a more well-mixed atmosphere that can transfer angular momentum from the deep interior.

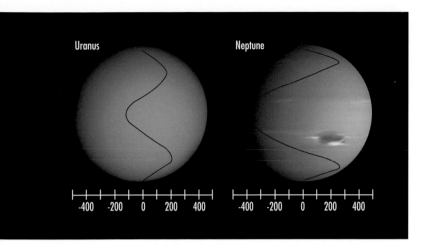

of helium and methane toward the core in Saturn and Neptune). For unknown reasons, Uranus experiences minimal internal heating and thus the weakest winds. Distant Neptune gets the largest fraction of its energy from within.

The length of day on a planet also radically affects atmospheric dynamics. In rapidly rotating planets like Jupiter and Saturn, the thick atmosphere separates into distinct columns of fluid rotating in different directions. The expression of these columns at the cloud tops is the alternating sequence of narrow jets and bands that we observe. More slowly rotating planets like Uranus and Neptune are unable to maintain distinct columns and instead develop a more well-mixed atmosphere. Because of this mixing, angular momentum

Giant Planets and Earth: Energy, Rotation, Winds				
Planet	Distance from Sun (AU)	Ratio of Energy Out to Sunlight In	Length of Day (hours)	Fastest Winds (m/s)
Earth	1.0	1.0	24.0	140 (in a tornado)
Jupiter	5.2	1.7	9.9	190
Saturn	9.5	1.8	10.6	395
Uranus	19.2	1.1	17.2	160
Neptune	30.1	2.6	16.1	600

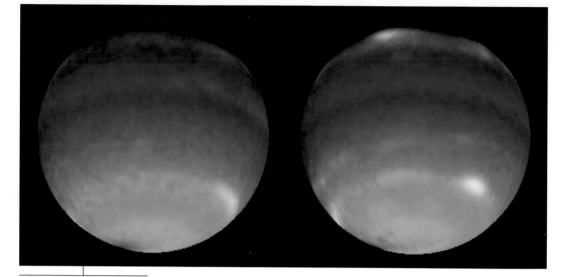

Hubble Space Telescope images show Neptune's blustery weather on both sides of the planet in 1996. The equatorial jet is depicted in deep blue in these false-color images. White, yellow, and red indicate high clouds above the blue methane layer. The green band encircling the pole is due to blue light being absorbed, possibly by atmospheric haze.

migrates from the faster-moving equatorial regions toward the slower poles and vice versa. The net result: strong westward winds surge at the equator and strong eastward jets flow at higher latitudes.

So Neptune appears to have the perfect combination of many factors—low friction, large internal heat, slow rotation that concentrates westward winds near the equator, and enough turbulence to develop large storms like the Great Dark Spot. Voilà! Our coldest major planet has the most wicked winds in the Solar System.

Best Vacuum Cleaner—Martian Dust Devils

Dust devils observed by the Mars Exploration Rover Spirit in Gusev Crater on August 7, 2005.

It is a dry, dusty day—not much different from any other day on Mars. Opportunity, one of two rovers exploring the martian surface, moves slowly across the landscape. The rover has been trekking across the martian desert for weeks now, covered in dust and occasionally getting bogged down by piles of sand. Progress is slow, and Opportunity's batteries are running desperately low.

A swirling mass of dust appears in the distance. Deceptively small at first, the spinning column looms larger as it approaches. It is coming straight toward the sluggish rover. As it hits, the swirling winds deftly wipe dust from the rover's solar panels. Opportunity's batteries begin recharging, giving the rover a new lease on life. The rover keeps on going and going.

Luckily for both Mars Exploration Rovers Opportunity and Spirit, this process happens again and again. Indeed, the vacuum cleaner action of martian dust devils has helped extend the historic missions of these rovers from a nominal 90 days to over six years (and counting)!

Dust devils usually form in dry, warm regions on Earth and Mars. Most often, dust devils arise from strong surface heating. Warm air rises and begins to spin as cooler air spirals in from all directions. Powerful vortices can also spawn from turbulent eddies created by gusty winds or surface obstacles. These swirling winds, if strong

Martian dust devil captured from above by the Mars Global Surveyor (MGS) Mars Orbiter Camera (MOC) in December 2002. The dust devil, nearly 3 km (1.9 miles) tall, left a surface streak behind as it moved from upper left to lower right. The shadow indicates a transition from a tight vortex near the surface to a broader dust plume at higher altitudes.

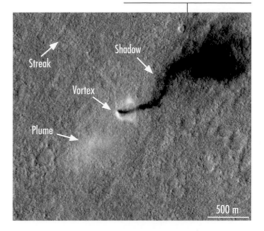

Streak

Shadow

Vortex

Plume

500 m

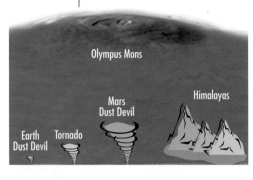

Martian dust devils are significantly larger than dust devils and tornadoes on Earth. The largest martian storms are as high as Mount Everest.

Scientists in Arizona study electrical and magnetic activity in dust devils as part of the Martian Atmosphere and Dust in the Optical and Radio (MATADOR) field experiment.

enough, lift fine dust particles into the air; without the dust, the spinning column of air would be largely invisible. Dust devils move in the general direction of the background winds, like whirlpools traveling downstream in a river. They move in straight lines when pushed by strong background winds or in loops during relatively calm conditions.

Most dust devils on Earth are relatively benign, small whirling dervishes that cause little damage. You may even have been hit by a dust devil with little to no impact. (We admit it—we've chased after dust devils just to get inside them.) But a few terrestrial dust devils are much more menacing. In the great desert regions on Earth, dust devils can get as tall as 10-story buildings (40 m/130 ft high), with freak dust devils reported as high as 2.4 km (1.5 miles) in Australia.

Nevertheless, martian dust devils dwarf those on Earth. The largest martian dust devils can be over 1.6 km (1.0 mile) wide and taller than Mount Everest. Winds whip around these huge vortices at over 90 m/s (200 mph). Even an average martian dust devil can make typical tornadoes on Earth seem meager in comparison. The huge size of martian dust devils is caused largely by lower values of gravity (38% of Earth's) and low atmospheric pressure (only 0.6% of Earth's).

Yet even more amazing, these monster storms are really gentle giants. Because of the thin martian atmosphere, the fast winds don't pack much of a punch. The Opportunity and Spirit rovers have suffered no damage from encounters with dust devils.

These devilish vortices move a lot of dust. Given the low atmospheric pressures on Mars, near hurricane-force winds are required to lift even small grains of dust. As dust devils churn across the surface, they sweep away a thin layer of dust and expose the darker underlying substrate. The dark tracks left behind often remain for a month or more. Dust devil tracks are found almost everywhere on

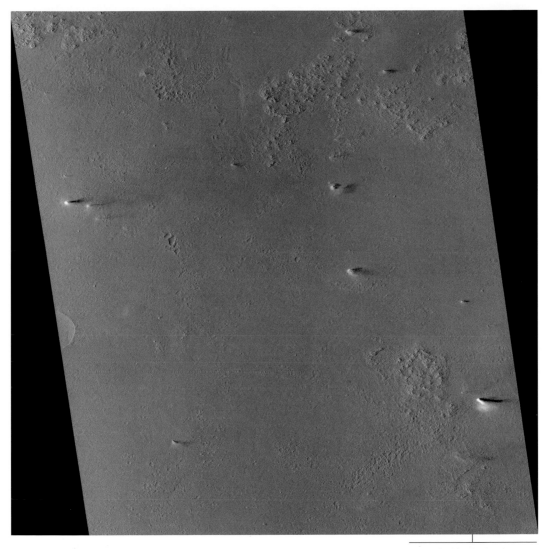

An armada of dust devils in the Amazonis Planitia region, a popular spawning ground for martian dust devil activity, as detected by MRO in November 2006. This image spans 30 km (19 miles) across. Dust is raised somewhere on the planet nearly every day.

Mars, from the deep Hellas Basin to high atop the Tharsis volcanoes. They have even been observed within the martian Arctic Circle, perhaps caused by dust devils whipped up by turbulent winds off the ice cap.

Consistent with formation by surface heating, dust devil activity peaks in the midsummer. However, dust devils have been observed

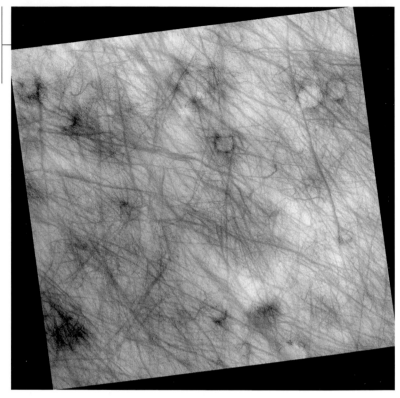

throughout the martian year. Frequent dust devils may explain seasonal surface darkening in the northern plains of Mars that we've been observing from here on Earth for over 100 years. They may essentially perform intense "spring cleaning" of the surface.

It is a cleaning job not to be taken lightly. After all, the surface of Mars is completely covered in dust. It takes a special combination—taller-than-Everest sizes, hurricane-force winds, and a large crew (many devils)—to do the job. Martian dust devils are, without a doubt, the most extreme vacuum cleaners in the Solar System.

The Hardest Rain—Diamond Hail on Uranus and Neptune

Ping, ping, ping-ping, ping. Thunk-thunk, thunk-thunk-thunk, thunk, thunk-thunk. Bam, bam, BAM!

Ah, yes, the beautiful sounds of hailstones hitting your car. Falling at speeds as high as 50 m/s (110 mph), hail can inflict serious damage to the windshield of your car or rooftop of your house. Annually, hail causes over $1 billion in crop damage and roughly $1.5 billion in property damage in the United States alone.

But it could be much worse. Fortunately, water ice is a relatively soft material. Depending on temperature, it ranges from 1.5 to 6 on the Mohs scale of hardness. Since hail typically forms at *relatively* warm temperatures between 0°C (32°F) and –30°C (–22°F), the hardness of hail is only about 2, slightly less than that of a human fingernail (this is why you can dig into a hailstone with your fingernail). Imagine the damage that would be caused by much harder material, such as pellets of solid iron, sleet of quartz crystals, or perhaps hail made of the hardest natural material of all: diamond.

Composite image of Uranus (left) in 2006 from the Hubble Space Telescope. The box highlights a dark spot, the first to be detected on the ice giant. Storm systems and dark spots on Neptune (right) as seen by Voyager 2 in 1989. The blue color of both planets is caused by methane absorption of red and green wavelengths. Methane deep in the interior of the planets may be compressed into diamonds, which may rain out onto the solid core.

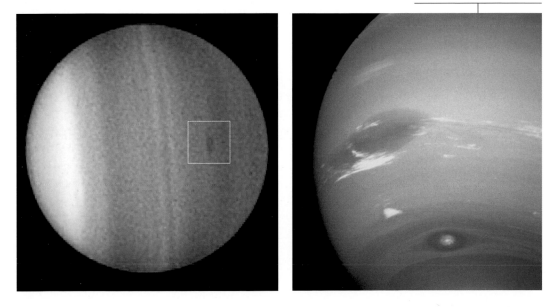

Hardness of Different Materials	
Material	Mohs Hardness
Graphite (carbon)	1
Ice (0°C/32°F)	1.5
Hail	1.5–2.5
Fingernail	2.5
Gold	2.5–3
Iron	4–5
Ice (−70°C/−94°F)	6
Glass	6–7
Quartz	7
Steel	7–8
Diamond (carbon)	10

The Mohs hardness scale is nonlinear: diamond is 1,500 times harder than graphite.

This may be exactly what happens on the ice giants Uranus and Neptune—diamond hail raining toward the center of these planets.

Uranus and Neptune are different from the other giant planets. Whereas Jupiter and Saturn are over 99% hydrogen and helium (with a smattering of nitrogen, oxygen, and carbon compounds for good color), Uranus and Neptune contain mostly "ices" of water, ammonia, and methane; only 15% of Uranus and Neptune is made of hydrogen and helium. In fact, Uranus and Neptune are closer in composition to the icy satellites of Jupiter and Saturn than to the gas giant planets themselves!

The exact cause of these differences is unclear, but the most likely explanation involves the sizes of the planetary cores in the early stages of Solar System formation. The gas giants (Jupiter and Saturn) had large enough rocky cores to gravitationally trap the lightest elements hydrogen and helium, whereas the ice giants (Uranus and Neptune), with smaller cores, could only hold on to heavier molecules. As the ice giants grew in size, they captured some remaining bits of hydrogen and helium to create an outer shell of atmosphere. Methane gas bubbled up from below to give Uranus and Neptune their characteristic blue-green hues.

But the atmospheres of these distant outer planets are far too cold to produce diamonds. For this, we must travel deep into their interiors. A thick mantle of icy slush—water, ammonia, and methane—takes up most of the volume of the ice giants. High temperatures within the mantle likely break up methane into its hydrogen and carbon components. Then intense pressures possibly squeeze this free carbon into a crystalline lattice that gives diamond its superior hardness. Diamond hail, as small as salt grains or as large as boulders, may steadily rain through the liquid mantle and pummel the rocky core. The core may be covered in a thick layer of diamonds, more massive than any diamond mine on Earth.

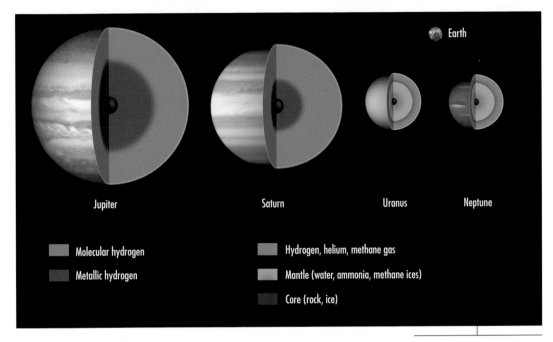

Jupiter Saturn Uranus Neptune

Molecular hydrogen

Metallic hydrogen

Hydrogen, helium, methane gas

Mantle (water, ammonia, methane ices)

Core (rock, ice)

The interiors of the giant planets all have rocky ice cores, but the similarities end there. Jupiter and Saturn have thick layers of metallic hydrogen created by intense pressures and outer envelopes of molecular hydrogen and helium. In contrast, Uranus and Neptune have mantles of heavy ices surrounded by atmospheres of hydrogen, helium, and methane. Earth is shown here for scale.

Of course, no one has actually seen diamonds on Uranus or Neptune. In fact, there is contradictory evidence regarding whether such diamonds can even exist. Laboratory experiments show that liquid methane, when subjected to pressures and temperatures found within the mantles of the ice giants, partially transforms into black diamond "dust." (Scientists create these un-Earthly conditions in the laboratory using a diamond anvil cell to produce pressures 100,000–500,000 times greater than Earth's atmospheric pressure and a laser to raise the temperature above 1,700°C/3,100°F.) However, recent computer simulations suggest that although the ice giant planets have an abundance of methane, there is not enough carbon to make diamonds quickly. With only 1%–2% carbon content on these planets, the chemical process that forms diamond would possibly take longer than the age of the universe.

Without a doubt, we need more data to better understand the inner workings of Uranus and Neptune and resolve this "gem" of a controversy. Unfortunately, no spacecraft missions to the ice giants are

An anvil cell consisting of two diamond tips can squeeze material (such as liquid methane) to high pressures in order to study conditions inside the giant planets. Here, a laser vaporizes one of the tips, sending a shock wave that rapidly increases the temperature of the material. Liquid methane turns into a hydrocarbon soup with black diamond specks in these types of high-pressure experiments.

currently scheduled. One promising NASA proposal for a Neptune flyby would not launch until 2019 due to a shortage of plutonium to power the spacecraft, and even then the spacecraft would not arrive at Neptune until 2028. The hardest rains in the Solar System will be, well, hard to detect for a while longer.

Part Four

Extreme Climates

Biggest, Baddest Babies of Climate— El Niño and La Niña

Fish mysteriously disappear off the coast of Peru. Disastrous fires rage in the rainforests of Indonesia. Crops in southern Africa wilt in severe drought, while at the same time, fields in central California drown in torrential floods. It happens every two to seven years, usually peaking during the Christmas season. Peruvian fishermen call it El Niño after the Christ child. It is one of the most devastating climate shifts in the Solar System.

Along with its baby sister La Niña, El Niño is part of a powerful ocean-atmosphere phenomenon in the equatorial Pacific Ocean. Called the El Niño–Southern Oscillation (ENSO), this climatic disruption involves distorted ocean currents, drifting warm pools of water, and dramatic shifts in atmospheric pressure and winds.

Historically, the terms *El Niño* and *La Niña* refer to oceanic warming or cooling in the eastern Pacific, while *Southern Oscillation* corresponds to fluctuations in sea-level atmospheric pressure in the western Pacific between Tahiti and Darwin, Australia. In the past few decades, scientists have discovered that these two events are inextricably coupled and have a striking impact on global weather and climate.

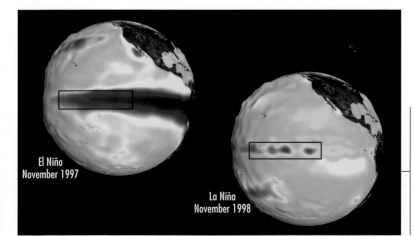

El Niño
November 1997

La Niña
November 1998

Sea surface temperatures in the Pacific Ocean for El Niño and La Niña conditions, with red being warmer than normal and blue colder than normal. Raised, highly exaggerated surfaces indicate sea level variations. The box in each image marks the Niño3.4 region (5°S–5°N, 170°W–120°W) used to monitor the intensity of El Niño and La Niña.

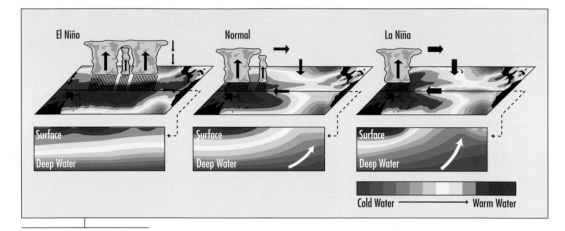

Ocean temperatures and atmospheric weather patterns in the equatorial Pacific Ocean. South America is to the right of each figure, and Australia-Indonesia to the left. El Niño is characterized by warmer surface waters in the eastern Pacific and a shift in rainfall to the east. Cold upwelling from the deep ocean is shut off by warm surface water. In contrast, La Niña exhibits cool eastern Pacific waters, stronger-than-normal trade winds, and heavy rain in the western Pacific. The atmospheric Walker circulation cell, apparent during normal and La Niña conditions, is absent during El Niño.

The "normal" state of the equatorial Pacific evolves like this: Trade winds continually blow from the east across the large expanse of ocean as a result of the Earth's rotation. These winds push surface water toward the west to produce a thick layer of warm water in the western Pacific. Cold, deep ocean water replaces warm water near the surface in the eastern Pacific, bringing with it nutrients (a respectable term for detritus) from the depths to support phytoplankton blooms. As the foundation of the oceanic food chain, these plankton blooms make the coasts of Ecuador and Peru among the premier fishing grounds in the world.

All of this warm water in the western Pacific serves as a powerful heat source for intense weather systems. A persistent low-pressure system forms in the west—and with it comes a lot of rain. Four-month-long Indonesian monsoons produce dense rainforests, and the tallest thunderstorms in the world occur just off the coast of Darwin. At the top of the troposphere, winds return eastward to complete a Pacific-wide circulation pattern called the Walker circulation (named after the discoverer of the Southern Oscillation). Fair weather reigns in the eastern Pacific as cloud-free air descends to the surface.

But things don't stay "normal" for long. Despite its name, the Pacific Ocean is anything but calm.

Think of the Pacific Ocean basin as a giant bathtub. As winds blow from the east, warm water begins to pile up to the west—average sea level can be almost 25 cm (12 in) higher in the western Pacific. This massive heap of water can't last forever—after a while, water in the big Pacific bathtub begins sloshing back. The warm pool of water migrates eastward, pushing the atmosphere with it. Opposing trade winds diminish, atmospheric pressure drops in the central Pacific, and rain moves away from Indonesia.

High surf and heavy rains eroded the California coastline during the 1997–98 El Niño, one of the strongest on record.

This is El Niño. Cold upwelling from the deep ocean is now blocked by a thicker warm surface layer in the eastern Pacific. With 60% fewer nutrients in Peruvian waters, fish go elsewhere in search of food. Seabirds and sea lions go hungry, and the famed Galapagos iguanas have mortality rates as high as 90%. Land-clearing fires, set by plantation owners in Indonesia and normally extinguished by monsoonal downpours, burn uncontrollably.

And then the Pacific Ocean sloshes back to the west. Trade winds strengthen, and the warm surface water moves back toward Indonesia and Australia. But the water overshoots the "normal" location and piles up even higher in the western Pacific. In the east, surface water becomes cooler than normal and the thermocline—the boundary between warm surface water and cold deep water—becomes very shallow.

This is La Niña, the "cold" phase of ENSO. (La Niña was initially called the anti-El Niño, but this translates as the "anti-Christ.") Heavy flooding ensues in Indonesia and Australia, excessive drought dominates South America, and life flourishes in the Galapagos.

The effects of ENSO extend well beyond the Pacific equatorial regions. The moving mass of warm water influences climate patterns worldwide—it deflects jet streams and moves weather to locations where it normally doesn't occur. Devastating drought and famine in Ethiopia have been linked to the intense 1982–83 El Niño,

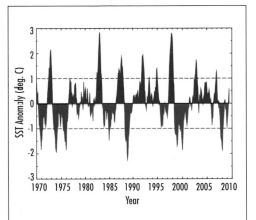

The Pacific Ocean oscillates in an irregular pattern between El Niño and La Niña every two to seven years. Originally, four regions in the Pacific were used to monitor ENSO. In 2003, scientists recognized that sea surface temperatures (SST) in a partial combination of regions 3 and 4 (aptly called Niño3.4) best captured widespread global climate variability. This plot shows the average weekly SST anomaly in the Niño3.4 region over the past 40 years. Values greater than 0.5°C indicate El Niño; values less than –0.5°C indicate La Niña. Niño3.4 index values exceeding ±1°C depict strong ENSO phases.

El Niño and La Niña alter the normal winter weather patterns in North America by modifying the location of the jet streams.

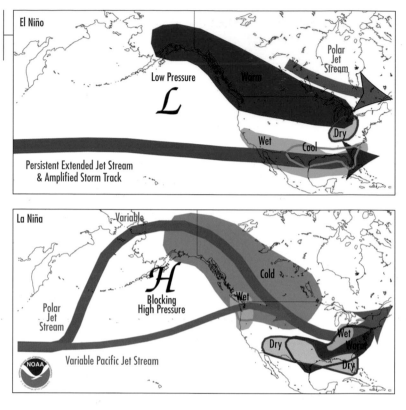

and record flooding ripped through central Europe during the 1997–98 El Niño. La Niña is equally disruptive on a global scale: stronger and more frequent Atlantic hurricanes, intense blizzards in central North America, and killer floods in Bangladesh.

Next to the tilt of the Earth's axis that produces our seasons, ENSO has the single largest impact on Earth's short-term climate. No other planet in the Solar System has anything like it. El Niño and La Niña are the biggest, baddest climate babies on the planetary block.

But It's a Dry Heat—Runaway Greenhouse of Venus

RUN AWAY! THE GREENHOUSE EFFECT IS CONTROLLING OUR PLANET!!!

A SMELT-ER-ING SUMMER—HOT ENOUGH TO MELT LEAD!!!

HELLO GLOBAL WARMING, GOOD-BYE OCEANS!!!

These alarmist headlines would make any rational person (and scientists *usually* fall into this category) skeptical about climate change and impending doom. Although such pronouncements may not make much sense for our Earth, they distinctly do apply to the planet Venus.

Don't get us wrong—the climate change issue on Earth is extremely important. After all, 2005 was the warmest year on instrumental

Ultraviolet image of Venus captured by the Pioneer Venus spacecraft in 1979. Thick clouds obscure the planet's sweltering surface caused by a runaway greenhouse effect.

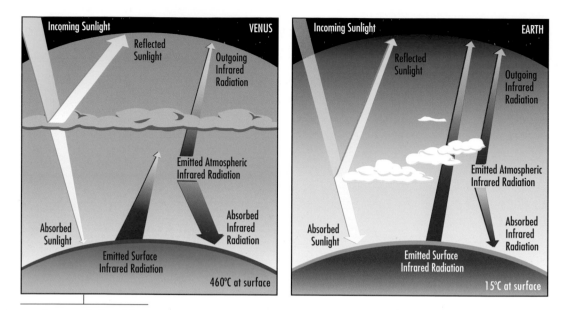

Comparison of radiation balance and greenhouse effects on Venus and Earth. Less sunlight reaches the surface of Venus, but the thick CO_2 atmosphere traps large amounts of infrared radiation near the surface. Venus's greenhouse effect is roughly 10 times stronger than Earth's.

record (since 1850). Based on tree ring studies, it was perhaps the warmest year for the past two millennia. Analysis of ice cores from Antarctica show that atmospheric carbon dioxide (CO_2), a strong contributor to global warming, reached historic levels in 2008—the highest amount in the last 650,000 years. The Intergovernmental Panel on Climate Change (IPCC) reported in 2007 that "warming of the climate system is unequivocal" and that humans were the primary cause of this increase through fossil fuel burning and land-use change (with 90% certainty—a remarkably emphatic statistic considering the complexity of the Earth system). Clearly global warming is happening on Earth.

It's just that, compared to Earth, the climate on Venus is in over-drive. Often called our sister planet, Venus is roughly the same size, density, and composition as Earth. Because of thick clouds that enshroud the entire planet, Venus was once thought to harbor steamy jungles teeming with exotic life.

You might expect Venus to be warmer than Earth; after all, Venus is 42 million km (26 million miles) closer to the Sun. However, the Sun is not the only thing controlling the Venusian climate. Thick sulfu-ric acid clouds reflect 80% of incoming solar radiation—*less* sunlight actually makes it to the surface of Venus than reaches the surface of Earth. Venus's surface temperature should be an icy 57°C (135°F) *cooler* than Earth's.

Yet observations show that just the opposite is true. Venus's surface temperature is a sizzling 460°C (860°F)—hot enough to melt lead! But it's a dry heat—any water that may have been on the surface boiled off long ago. These searing temperatures can't be due to sunlight alone. Surprisingly, the temperatures are relatively uniform from equator to pole *and* from day to night. It's as if Venus is covered with a warm blanket that keeps the entire planet toasty. This blanket is actually an atmosphere full of carbon dioxide and it produces a runaway greenhouse effect on Venus.

Global warming occurs when the delicate balance of incoming and outgoing radiation is disturbed, and greenhouse gases like carbon dioxide can alter this balance. Greenhouse gases do not absorb much radiation at visible wavelengths (i.e., there are visible "windows"), but they are great absorbers at various infrared wavelengths. They do allow some infrared radiation to escape to space within "infrared windows"; otherwise planets would have a very difficult time cooling off.

The greenhouse process happens like this: Solar radiation reaches the surface through a visible "window" and warms the surface. In

Although early-20th-century scientists and science fiction writers suggested the possibility of bizarre life on Venus, there is no evidence of flying dinosaurs on the swampless planet. But it is hot. . . .

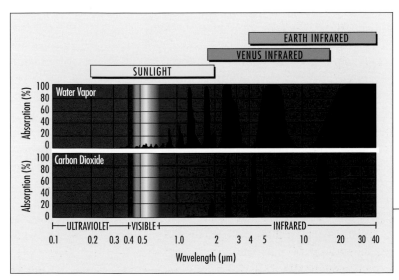

Because of their different temperatures, the Sun, Venus, and Earth emit radiation at different peak wavelengths. Water vapor and carbon dioxide are considered greenhouse gases because of their strong absorption in the infrared. Water vapor is the primary greenhouse gas on Earth, whereas carbon dioxide controls the climate on Venus.

One of the few visible images of the warmest planetary surface in the Solar System. The Venera 13 lander survived on the surface for only two hours and seven minutes in March 1982. Part of the spacecraft is visible at the bottom of this image.

an attempt to cool, the warm planetary surface emits energy in the infrared. Before this infrared radiation can make it to space, however, greenhouse gases absorb it and then reemit infrared radiation both upward and downward. Upward infrared radiation eventually does make it to space, but downward infrared radiation heads back to the surface and warms it even more. In essence, more energy (in the form of radiation) reaches the surface than just from the Sun alone.

This infrared radiation from the atmosphere is potent. On average, Earth's surface receives 88% more radiation from the atmosphere and clouds than from the Sun! The result is even more extreme on Venus—the atmosphere provides almost 1,000 times more radiation at the surface. Without the greenhouse effect, surface temperatures of both Earth and Venus would be below freezing.

Why is the greenhouse effect so strong on Venus? The Venusian atmosphere is 90 times more massive than the Earth's atmosphere, exerting a surface pressure equivalent to that experienced at an ocean depth of 910 m (3,000 ft). In addition, 96% of this thick atmosphere is carbon dioxide. Whereas most carbon on Earth is sequestered in rocks, most carbon on Venus resides in the atmosphere. High pressures cause carbon dioxide to be more effective at absorbing infrared radiation (the infrared windows that let heat escape become narrower). An overabundance of available carbon dioxide results in a runaway greenhouse effect—little infrared radiation from the surface escapes to space.

With a thick, suffocating atmosphere and temperatures higher than a broiling oven, Venus has the hottest surface of any planet in our Solar System. Global warming and the greenhouse effect are alarmingly real. Just ask Venus.

Dirtiest Climate Change—Global Dust Storms on Mars

In 1971, NASA sent the Mariner 9 spacecraft to Mars to address some of the most compelling questions in space exploration: What does the martian surface look like? Are there indeed canals on Mars, as first suggested by Schiaparelli in 1877? Does an advanced civilization live on the Red Planet? Earlier spacecraft had flown by Mars and imaged small fractions of the planet, but Mariner 9 would be the first spacecraft to orbit another planet. With its state-of-the-art camera, it would image nearly 80% of Mars and offer definitive evidence on the state of the surface.

To the dismay of planetary scientists, Mariner 9 saw practically nothing of the martian surface when it arrived in November 1971. A global dust storm raged across the planet and almost completely obscured surface features. Luckily, NASA was able to reprogram Mariner 9 to wait out the storm. When the dust finally settled in

Hubble Space Telescope images capture a global dust storm on Mars in 2001. The left image shows a small regional dust storm at the lower right of Mars. A few months later, the dust storm covered the whole planet.

June 26, 2001   September 4, 2001

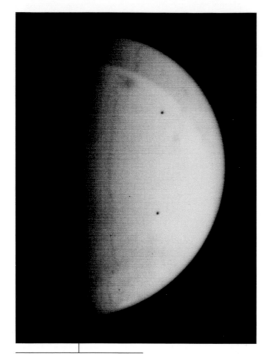

Image of Mars in 1971 as Mariner 9 approached the planet. A global dust storm enshrouds the planet. The fuzzy dark spot at the upper edge is Olympus Mons, the tallest mountain in the Solar System, whose summit extends higher than the dust storm. The more distinct dark spots in the image are calibration guides on the camera lens, and the ghost image is an artifact of the image recording process (this problem was fixed for the next spacecraft, Mariner 10).

January 1972, Mariner 9 made some remarkable discoveries—the tallest volcano, Olympus Mons; the deepest canyon, Valles Marineris (named after Mariner 9); and extensive dry river channels that suggested flowing water in the distant past. There was no evidence of a canal system developed by advanced life-forms.

Yet as often happens in science, initial problems provide unforeseen opportunities. Mariner 9 inspired scientists to better understand global dust storms on Mars. As it turns out, martian global dust storms are one of the most extreme agents of climate change in the Solar System.

Mars is a huge desert—drier than any desert on Earth—with large sand dunes, dry ancient river channels, and timeworn mountain ridges. Blanketing this desolate landscape is a thin layer of red iron-rich dust, usually only a few millimeters thick but in some places nearly 2 m (6.6 ft) deep. It's a bit "iron-ic" that the infamous red color of Mars, the source of a mythological Chinese "fire star" warrior and of the bloody Roman god of war, is actually caused by a very thin layer of rusty iron specks.

So with all of this dust around, you might think that global dust storms are the norm on Mars. Local dust devils and small regional dust storms do occur regularly, but global dust storms are much less frequent. Since 1956, only eight giant dust storms have encircled the entire planet, the most recent in 2007. Remarkably, they happen only during the southern hemisphere's spring and summer seasons.

It takes a special combination of ingredients to get a global dust storm. First, extremely strong *horizontal* winds (20–30 m/s or 45–67 mph) are needed to lift dust into the atmosphere. The martian atmosphere is only 0.6% as dense as Earth's. Fewer air molecules hit the martian surface, so each molecule needs a lot of kinetic energy to get dust into the air. Second, these winds must be persistent. Dust devils and small regional dust storms that last only a few hours or days have little chance of stirring up enough dust to cover the planet. Third, *vertical* winds must be vigorous to inject copious amounts of dust high into the stratosphere. Global circulation patterns in the upper atmosphere then can spread dust around the entire planet.

How are such strong, persistent winds created on Mars? One way is by intense temperature changes that occur in springtime near the polar region—the stronger the temperature difference, the stronger the winds. Cold, dense air plunges off the polar ice cap and rushes underneath warmer air in surrounding ice-free regions. In a similar fashion, cold air at high elevations flows down relatively steep slopes into warmer basins—the 2001 global dust storm originated deep in Hellas Basin, a large impact structure in the Red Planet's southern hemisphere. And since the egg-shaped orbit of Mars places it closer to the Sun in southern spring and summer, temperature differences become more dramatic during these seasons, thereby producing stronger winds.

When global dust storms do occur, they produce rapid global warming in the atmosphere and intense cooling at the surface. Dust in the atmosphere absorbs incoming solar radiation and causes atmospheric temperatures to rise. On the ground, the skies appear much darker during a dust storm and the temperature drops dramatically. At the peak of the 2001 dust storm, upper *atmospheric* temperatures increased by 40°C (72°F) and the daytime global *surface* temperature dropped by 23°C (41°F) compared to the previous martian year.

But the effect on martian climate doesn't end when the dust storm is over. Previously darker areas around the planet are now covered with lighter-colored dust, and more sunlight is reflected away from the surface. For two years after the 2001 dust storm, global daytime surface temperatures remained 2–3°C (4–5°F) cooler than before. This is no minor climate change: this temperature decrease on Mars *within months* caused by dust is comparable in magnitude to the

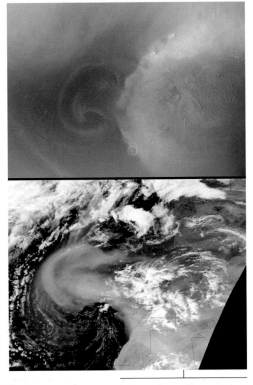

Intense winds caused by temperature differences stir up giant tongues of dust on Mars (top) and Earth (bottom). Cold winds off the Mars north polar seasonal ice cap blow dust roughly 900 km (560 miles) away from the cap edge. On Earth, similar circulations due to land-water temperature differences blast African dust roughly 1,800 km (1,100 miles) over the Atlantic Ocean.

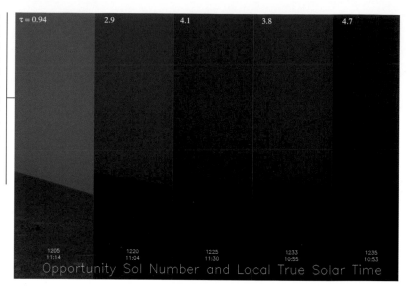

Dark days for the Opportunity rover at Victoria Crater, Mars. A regional dust storm severely darkens the skies in martian summer 2007. A *sol* is a martian day, roughly 40 minutes longer than an Earth day. τ represents the optical thickness of the dust (clearer skies have lower numbers), but the effect on sunlight is exponential: dust blocks out over 99% of direct sunlight by Sol 1235. With reduced sunlight, the surface temperature plummets.

global temperature increase predicted in the next *fifty years* due to enhanced greenhouse gases on Earth.

So imagine this scenario on Earth: A roiling dust storm quickly covers the entire planet, blocks out the Sun, and plunges warm springtime temperatures into a deep freeze. Dust fills the air, polluting the skies with a dingy red tinge. Even after the skies clear, temperatures remain significantly cooler for years. It seems so unlikely on Earth, but it happens every few years on Mars. It is the dustiest, dirtiest example of climate change anywhere in the Solar System.

Most Bizarre Seasons—Uranus

Imagine, for a moment, that you live on an Earth with a wildly skewed climate. You stand at the North Pole in the middle of summer, basking in tropical warmth. The Sun never sets—it stays in roughly the same location directly overhead, day after day. In fact, you have a hard time distinguishing when one sweltering day ends and another begins. Meanwhile, your friend in Ecuador—near the equator—barely sees the Sun at all. It skims the horizon in a 360° circuit, and the tropics remain in perpetual dusk. On the other side of the planet, Antarctica and most of the southern hemisphere stare into the cold darkness of deep space . . . a long, uninterrupted winter's night.

The poles and the tropics switching roles, half of the planet stuck either in continual light or unending darkness—such things sound too bizarre to be possibly true. But they actually happen on the planet Uranus.

Uranus remains one of the least understood planets in our Solar System. Although discovered by Sir William Herschel in 1781

Uranus in true color (left) and false color (right) as seen by the Voyager 2 spacecraft in 1986. The south pole is just slightly off center in these images. Fuzziness along the right edge of the true-color image marks the day-night terminator, beyond which the northern hemisphere lies in total darkness. The false-color image is produced from a combination of ultraviolet, violet, and orange light, and extreme contrast enhancement brings out atmospheric features near the poles. Regions in darkness (such as beyond the terminator) appear white in the false-color image.

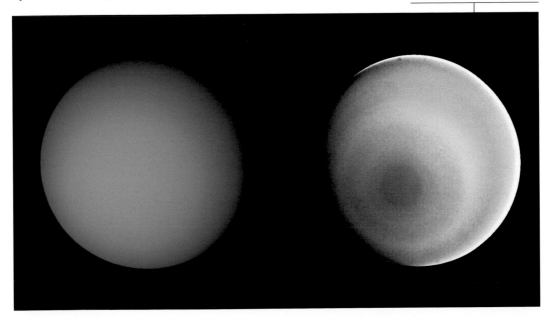

Mercury	Venus	Earth	Mars	Jupiter
0.1°	177°	23°	25°	3°

The obliquity (axial tilt) of the eight major planets. A larger tilt leads to more pronounced seasonal effects. Uranus's 98° obliquity produces intense summers and dark winters at the poles.

(Herschel originally thought Uranus was a comet), detailed observations of the distant planet were scarce until the late 20th century. When Voyager 2 cruised by Uranus in 1986, the spacecraft witnessed a relatively bland planet whose blue-green, methane-laden atmosphere lacked the turbulent storm systems found on the other giant planets. In fact, the Voyager Imaging Team needed to employ extreme image processing to discern any atmospheric structure at all.

This giant planet is anything but boring, however. Multiple rings made primarily of cobble- to boulder-sized particles encircle Uranus, while the orbits of its small irregular moons chaotically interact with one another—enough that a collision between moons is likely. But the most striking characteristic just might be that the planet seems to be tipped over on its side. The rotational axis is tilted 97.9° with respect to its orbital plane. (An axial tilt, or obliquity, of greater than 90° means that the south pole actually points northward.) Rather than spinning like a top as most other planets do, Uranus rolls around on its side as it makes its journey around the Sun. A massive collision with another planetary body in the early history of the Solar System likely caused Uranus to topple over. This exceptional tilt wreaks havoc on the Uranian seasons.

A planet's seasons are caused by its obliquity—the larger the tilt, the more extreme the seasons. Jupiter, for example, shows almost no

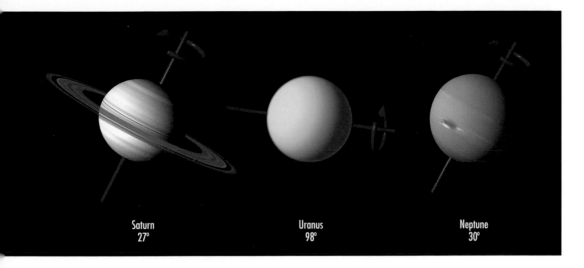

Saturn
27°

Uranus
98°

Neptune
30°

seasonal variation due to its nearly vertical axis. As a planet moves in its orbit, different parts of the planet are tilted toward the Sun. Earth's moderate tilt (23.4°) produces a relatively temperate northern summer when the North Pole is oriented toward the Sun. Six months later, a somewhat mild northern winter results when Earth's North Pole tilts away from the Sun. (Residents of the Arctic Circle may beg to differ, however.)

The process is even more dramatic on Uranus. Because of the large obliquity, one pole points almost directly at the Sun during summer

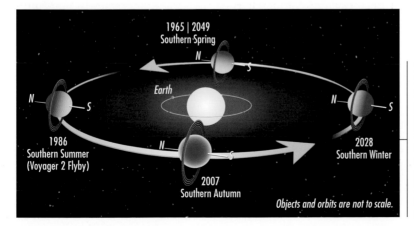

1965 | 2049
Southern Spring
N

Earth

N S

N S

1986
Southern Summer
(Voyager 2 Flyby)

N

2028
Southern Winter

2007
Southern Autumn

Objects and orbits are not to scale.

The almost horizontal orientation of Uranus's axis caused the south pole to directly face the Sun (southern summer) during Voyager 2's encounter in 1986. Twenty-one Earth years later, Uranus experienced southern autumn as the Sun moved directly over the equator. Southern winter will culminate in 2028 as the south pole faces deep space. Southern spring equinox, which last occurred in 1965, will return again in 2049.

Adaptive optics on Earth-based telescopes remove fuzziness caused by atmospheric turbulence and produce some of the best images of Uranus to date. These Keck Observatory near-infrared images at two different wavelengths show the power of this technique: the left set of images has adaptive optics turned off, while the right set has adaptive optics on. The bright dot in the upper images is Uranus's moon Miranda. The rings are excluded in the magnified lower images. South is to the left and slightly up in these images. Note the bright polar collar around the south pole and bright storm clouds in the northern hemisphere.

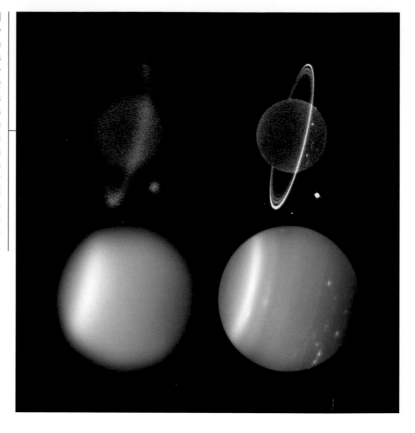

while the other pole points toward the dark outer reaches of the Solar System. Since it takes 84 years (longer than the average human life expectancy) for Uranus to orbit the Sun, this intense summer lasts about 21 years!* Luckily, autumn and spring are a bit more "normal." The Sun is directly over the equator and the entire planet experiences both day and night. Nevertheless, when averaged over an entire Uranian year, the poles receive more than twice the sunlight that falls on the equator.

At first glance, the bizarre seasons of Uranus are strangely benign. Unlike the other giant planets, Uranus receives most of its energy from the Sun rather than from internal heating. Simple theories

*Unless otherwise stated, a year refers to an Earth year = 365.24 Earth days.

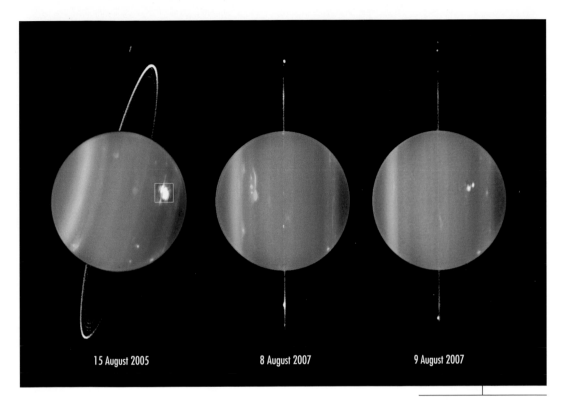

15 August 2005   8 August 2007   9 August 2007

based on sunlight absorption predict that the summer pole should be roughly 6°C warmer than the equator. Yet temperatures from pole to equator are surprisingly constant. Weather systems on Uranus must be incredibly efficient at moving energy from poles to equator to be able to maintain such uniform conditions, but evidence for such strong systems has remained notably absent.

Absent until now, that is. A new generation of large ground-based telescopes and advanced imaging techniques have allowed astronomers to detect storms that were invisible to Voyager 2's cameras (if such storms were even present during the Voyager encounter). And to make conditions even more favorable, Uranus is in the process of transitioning into a new climate. During Voyager's 1986 visit, Uranus was in southern summer, but southern autumnal equinox (the beginning of fall) recently occurred in 2007. The northern hemisphere is seeing the Sun for the first time in 42 years! Accordingly, weather in

Near-infrared images from the Keck II telescope on Mauna Kea show increased weather activity on Uranus. The south pole is to the left and the equator directly below the rings in all images. Uranus reached equinox in 2007, so the rings appear edge-on. As Uranus transitions from southern summer to northern summer, sunlight returns to the northern hemisphere and storms become active in that hemisphere. The bright south polar ring has dimmed considerably in the past few years, while a new north polar band has brightened.

the northern atmosphere is becoming more active; huge springtime storms, a developing collar of polar clouds, and hurricane-like vortices are making appearances. At the same time, the southern hemisphere is beginning its slide into a long, frigid winter.

Observations over the next few years will continue to shape our view of Uranus. As recent studies have shown, weather and climate on the seventh planet from the Sun are not so dull after all—the cold northern hemisphere is waking from a 42-year hibernation and things are getting exciting. Uranus is now experiencing one of the most extreme seasonal climate changes in the Solar System.

Snowballs in Hell—Mercury

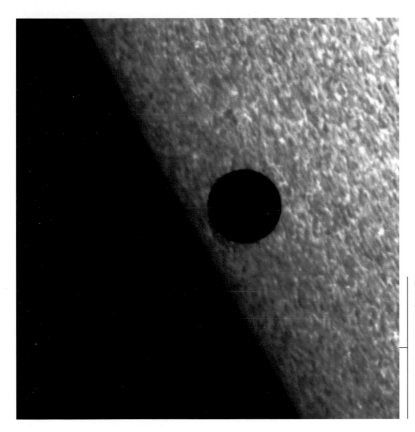

Mercury passing in front of the Sun on November 8, 2006, captured by the Solar Optical Telescope aboard Japan's Hinode satellite. The tiny terrestrial planet gets as close as 46 million km (28.5 million miles) from the Sun during the course of its eccentric orbit. Due to its proximity, Mercury can receive more than 10 times the maximum amount of solar radiation received at Earth.

The planet Mercury is hot. It's obvious, right? The tiny rocky body is only about 0.4 AU from the Sun on average—a mere 60 million km (36 million miles).* Noontime temperatures on the surface of our Solar System's innermost planet can exceed a sizzling 427°C (800°F). The average over the entire planet is more like 169°C (336°F)—not quite hot enough to melt lead but plenty hot enough to boil water! So it stands to reason that Mercury must be a sun-blasted, hellishly hot, barren wasteland. And it is . . . except, of course, for the ice.

---

*1 astronomical unit (AU) is the average distance between the Earth and the Sun.

The National Radio Astronomy Observatory's Very Large Array (VLA) in New Mexico (top), the Goldstone antenna (part of NASA's Deep Space Network; bottom left), and the radio telescope at the Arecibo Observatory in Puerto Rico (bottom right) were all used to detect ice on Mercury.

You read that correctly. We said ice! On Mercury! Who would have thought? Not many people, in fact, before 1991. That's when radar astronomers turned the collective power of NASA's Goldstone 70-m radio telescope, the Very Large Array (VLA; a 27-antenna array of 25-m radio telescopes), and the giant 305-m Arecibo radio telescope toward tiny Mercury.

Despite scorching temperatures in the daytime, the dark side of Mercury—the side opposite the Sun—can be as frigid as −173°C (−280°F). That would be plenty cold enough for ice, if it weren't for Mercury's rotation. For centuries it was believed that the innermost planet's rotation must be tidally locked to the Sun. And it is, just not in the 1:1 (one rotation per one orbit) synchronous rotation that everyone expected. Instead, Mercury rotates three times for every two orbits.

The orbit of the little planet has a large eccentricity—it's more egg-shaped than circular. In fact, Mercury's eccentricity is the largest of the eight planets—an eccentricity value of 0.2056 for Mercury compared to Earth's 0.0167 (a perfect circle has an eccentricity of 0). This large eccentricity allowed Mercury to become tidally locked into a stable 3:2 rotational resonance rather than the expected 1:1 of synchronous rotation (like our own Moon and other large moons in the outer Solar System). This 3:2 rotational resonance means that the same face of the planet does *not* always face the Sun. Mercury has no permanently dark side.

Goldstone-VLA radar images of Mercury made in 1991 (top) and of Mars in 1988 (bottom). Bright radar backscatter (red) from the north pole of Mercury is similar to bright radar returns from Mars's polar caps.

If the cold dark side eventually sees the light of day and is heated up to 427°C (800°F), why do we think there might be ice? Well, the maximum temperature is just that, a maximum. Such high values occur only when the Sun is directly overhead (local noon) and only near the equator. Unlike Earth, Mercury's rotation axis is not appreciably tilted, so the small planet experiences very little in the way of seasonal variation. Because they never tilt toward (or away from) the Sun, the north and south poles see the Sun only very near to the horizon. There are no blazing overhead noontime suns at Mercury's poles.

And that is the key to finding ice on Mercury. Since the Sun is at a very shallow angle above the horizon, it is possible for the floor of a crater in the polar regions to be in perpetual shadows . . . and never feel the warmth of the Sun. The walls of the crater rims can be tall enough to block what little sunlight arrives at Mercury's poles. Because the crater floors are always in darkness, temperatures there may never exceed –173°C (–280°F). Now *that's* cold enough for ice to stick around without sublimating for billions of years.

Okay, you say, ice could *possibly* exist—but do we really think it's present? Remember those radar astronomers we mentioned earlier? They actually found *evidence* for such ice. Bouncing radar waves off Mercury is one of the only ways to study the innermost planet from Earth. The signature of the returned radar signal is much more consistent with ice than with rock—the radar backscatter from

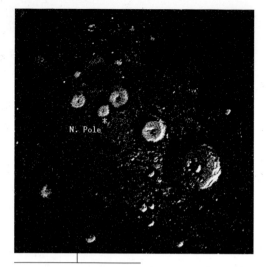

N. Pole

Because of favorable orientations between Earth and Mercury, the Arecibo telescope is able to look at the inner planet's polar regions. Radar-bright patches correspond to the floors of permanently shadowed craters, very suggestive of ice.

Mercury's pole strongly resembles that from Mars's polar ice caps.

The water could have been delivered to Mercury by impacting comets. Or perhaps the water originates from within Mercury itself— outgassing via volcanic eruptions could account for the trace amounts of water ice at the surface. The permanently shadowed craters would act as cold-traps and hold on to any water that made its way to the poles. Any water not in the shadows of a crater by sunrise, however, is doomed to sublimate away and be lost.

The bright radar reflection *could* be due to something other than water ice. One suggestion is condensed sulfur compounds, which may have radar backscatter properties similar to those of ice. So far, however, frozen water remains the simplest explanation. And we won't really know for sure until we are able to take a much closer look.

Fortunately, there are a couple of Mercury missions, either in operation or in planning, that will be able to help us determine the nature of the radar-bright polar deposits. NASA's MESSENGER spacecraft, the first to visit Mercury since Mariner 10 in 1975, has already completed three flybys of the planet and should settle into Mercury orbit in March 2011. Meanwhile, ESA's BepiColombo mission is in preparation for an August 2013 launch. With the suite of sophisticated instruments carried by these missions, we just might have more than a snowball's chance in hell of determining if there is ice at Mercury's poles.

Part Five

Rings and Things

Those Lovely Rings—Saturn

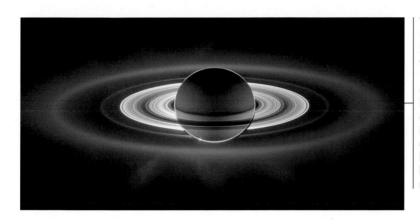

The rings of Saturn as seen by the Cassini spacecraft on September 15, 2006. This contrast-enhanced image was created from the combination of 165 images taken by the wide-angle camera over a period of about three hours. The Sun is hidden behind the disk of Saturn, allowing the rings to be illuminated from behind. This lighting angle revealed new, faint rings that had never before been seen. The pale blue dot just outside of the bright main rings at 10 o'clock is distant Earth.

Ah Saturn . . . and those amazing rings. Is there any other planet in our Solar System that is as immediately recognizable to us—other than perhaps our own blue Earth? The distinctive rings of Saturn stand out as the crown jewels of our Solar System. They have fascinated astronomers and planetary scientists since their discovery.

Although Saturn can be seen with the naked eye as a faint yellow wanderer in the night sky, the rings themselves can be seen only with the help of technology. Galileo Galilei, the father of modern astronomy, was one of the first scientists (if not *the* first) to turn the newly invented telescope skyward in an effort to understand the heavens.

By today's standards, Galileo's telescope was a rather poor one. It could magnify by only a factor of 20 (20x) or so. You can likely run to your nearest sporting goods shop and purchase a pair of moderately priced binoculars with better optical qualities for viewing the night sky (and we recommend you do so!).

Galileo's sketches of Saturn from 1610 (top) and 1616 (bottom).

However, in 1610 Galileo's state-of-the-art telescope revealed a remarkable and unexplainable feature: the planet Saturn had "ears." Even more baffling, these lobes would vanish and then reappear over the course of a few years.

It took 50 years before Dutch astronomer Christiaan Huygens—using a telescope twice as powerful as Galileo's—suggested that the "handles" were, in fact, a thin solid ring around the planet. Late in the

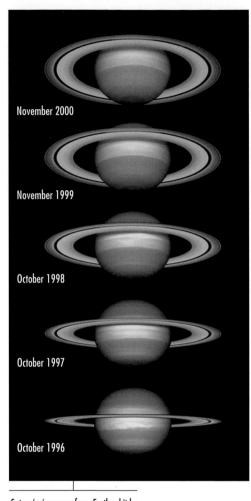

November 2000

November 1999

October 1998

October 1997

October 1996

Saturn's rings seen from Earth orbit by the Hubble Space Telescope (HST) at roughly yearly intervals. When nearly edge-on, as in the October 1996 view, the rings cannot be seen with smaller Earth-based telescopes such as the one used by Galileo in 1610. The view of the rings from Earth most recently became edge-on in September 2009.

17th century, Giovanni Cassini noticed a dark gap that suggested there were at least two such rings. In 1856, Scottish mathematician and physicist James Clerk Maxwell demonstrated that the rings must consist of numerous small particles rather than solid sheets, a natural result of Saturn's intense gravitational pull.

Yet after four centuries of study, our best understanding of Saturn and its rings has come only in recent decades. Spacecraft have obtained close-up views of the giant planet and its rings: Pioneer 11 (1979), Voyager 1 (1980), Voyager 2 (1981), and the aptly named Cassini-Huygens mission (2004–present). These robotic explorers have discovered that Saturn's rings are more complex and delicate than Galileo, Huygens, or Cassini could have ever imagined.

Saturn's ring system consists of seven major rings, named A through G in the order of discovery. The numerous ring particles, ranging in size from grains of sand to boulder-sized blocks as big as houses, are made of almost pure water ice. The main rings (A–C) are the brightest and most densely populated—is it any wonder that these were the only rings discovered using ground-based telescopes? The fainter D–G rings consist of tiny dust-sized ice particles and are best seen when Saturn is illuminated from behind by the Sun, an impossible view from the vantage point of Earth.

Although the rings at first appear smooth and regular, closer inspection reveals that they contain an amazingly complex structure. Funky things can be found in these rings—smaller ringlets, gaps, braided rings, clumpy ring arcs, scalloped edges, and spiral density waves. Many of these features are caused by critical gravitational interactions with the moons of Saturn.

One such interaction occurs when the orbits of ring particles sync up with the orbital motion of a moon. This orbital resonance is simi-

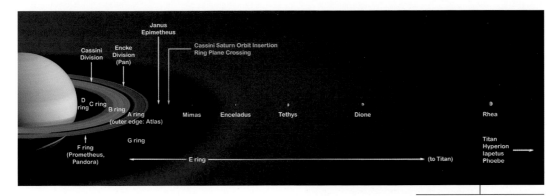

Saturn's rings in relation to some of its moons. The horizontal extent of the rings is vast—nearly 400,000 km (almost 250,000 miles), greater than the average distance from the Earth to the Moon! Yet the rings are remarkably thin, only a few tens of meters thick.

lar to pushing a child on a swing. If you time your pushes correctly, each little push adds up and the happy youngster swings higher and higher. Push at the wrong time, however, and you hinder rather than help the motion of the swing. The periodic orbits of Saturn's moons cause small gravitational tugs to be exerted on the ring particles with each pass. If a moon's orbit has the proper period, the tugs can add up to clear the icy chunks away, or they can hinder and keep the chunks confined.

One of the most prominent examples of resonance effects is the Cassini Division, which separates the A- and B-rings. Although it appears empty from a distance, the Cassini Division just has fewer and darker particles than the adjacent rings.

The orbit of Mimas is much farther from Saturn than the particles of the Cassini Division, but a powerful 2:1 resonance between the moon and ring particles ejects material from the dark divide. The ring particles must travel around Saturn twice for every one trip Mimas makes. Like getting a push on every other swing, the 2:1 resonance pushes the particles into more energetic orbits . . . and eventually out of the Cassini Division.

In contrast to Mimas, the moons Pandora and Prometheus are *shepherd moons* whose orbital tuggings confine ring particles rather than eject them. Like sheepdogs herding their woolly charges from

Like the wake produced by a boat moving through water, elaborate ring arcs, scalloped edges, and spiral density waves are produced by the small moon Pan as it orbits Saturn within the Encke Gap.

Pandora (left) and Prometheus (right) help keep the braided F-ring confined and produce much of the ring's complex structure.

Cryovolcanic geysers from the south polar regions of the moon Enceladus spew out water vapor and fine ice particles. This H_2O escapes the surface of the moon only to end up in orbit around Saturn and continually resupply the E-ring.

field to pen, the shepherd moons exert gravitational nips and nudges as they zip around Saturn, each moon going its own speed. These tugs combine to keep the small icy particles of the F-ring confined to . . . well, a ring. And because of the constant interactions with these moons, the narrow, braided F-ring is one of Saturn's most dynamically active rings.

But perhaps the most amazing recent discoveries of moon-ring interactions involve the moons Enceladus and Phoebe. These two moons have been caught in the act of ring building. Enceladus actively spews water vapor and ice crystals from its southern "Tiger Stripes" to replenish the diffuse E-ring. Meanwhile, the ice and dust being blasted from Phoebe's surface by meteoroid impacts create the largest and most diffuse ring yet discovered. Like the moon Phoebe, this immense, newly discovered ring is tilted 27° to Saturn's other rings and orbits in the opposite direction. Dust from this new ring peppers Saturn's bizarre moon Iapetus like bugs hitting a car windshield and may be responsible for that moon's yin-yang appearance.

Although impacts and geysers appear to account for the ice and dust in the tenuous outer rings, we still aren't sure how the main rings formed. The A-, B-, and C-rings contain roughly 20% more ice than in all of Earth's glaciers and ice sheets combined. Although only a small fraction of Earth's total, this is still a significant amount of water—over 450 times the mass of water in the Caspian Sea.

How did this much water find its way into a disk around Saturn in the first place? The ice chunks could have accreted into a disk along with Saturn as planets began forming within the solar nebula. Or, perhaps, close-passing comets were torn apart (like Comet Shoemaker-Levy 9 at Jupiter) and then captured into orbit around the gas giant. Maybe two (or more!) of Saturn's small icy moons collided, leaving icy debris

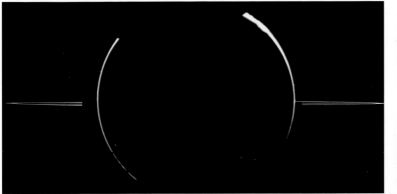

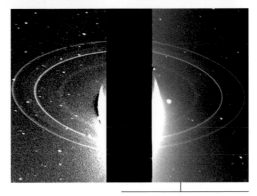

strewn about the region of their prior orbits. Instead of colliding moons, it's quite possible that a large moon (larger than Mimas) strayed too close to Saturn and was shredded by the gas giant's great tidal forces. The origin of planetary rings remains an unsolved mystery of planetary science. Even after 400 years, there is still much more to discover about Galileo's enigmatic "handles."

It may come as a surprise for some to learn that Saturn is not our only ringed planet. The other giant planets of our Solar System are also encircled by rings. Plus, recent measurements by the Cassini orbiter suggest that Saturn's moon Rhea may even have a very tenuous ring system of its own! However, these other meager ring systems pale in comparison to the magnificent set of rings that bejewel Saturn. So although Saturn may not have the only set of rings in our neighborhood, the sheer scale, brightness, and complexity of Saturn's rings make them the most iconic—and most lovely—rings in our Solar System.

The rings of the other giant planets: an edge-on view of Jupiter's rings as seen by the Galileo orbiter in 1996 (top left), the rings around Uranus as seen from Earth orbit by the Hubble Space Telescope (HST) in 2003 (top right), and Neptune's rings as seen by Voyager 2 in 1989 (bottom). These ring systems are more tenuous and much less bright than Saturn's amazing rings.

Billions and Billions of Bodies—The Oort Cloud

A schematic representation of the billions of icy bodies in the outer Solar System. The spherical outer Oort Cloud is only weakly bound by our Sun's gravitational attraction. The donut-shaped torus of the inner Oort Cloud — also called the Hills Cloud — probably contains 5–10 times more cometary nuclei than the larger, outer cloud. The zoomed view of the Kuiper Belt and the orbits of the outer planets illustrates the scale of the Oort Cloud with respect to more familiar parts of our Solar System.

Ever thought about just where comets come from—those weird streaks of fuzzy light that look like ghosts in the night sky? Your crazy uncle (you know, the one the family doesn't talk much about) would probably just tell you they come from a gigantic mystical cloud of ice balls—ethereal chunks of ice, maybe billions of them, that are just whizzing around out there in space. Of course, as is the case for many of his far-fetched stories, nobody has ever actually seen these myriad small icy bodies located far, far from the Sun. The way he tells it, those unearthly balls of ice must be so distant that the Sun can just barely hold on to them. Every so often something gives one of the balls of ice a little tweak and whoosh, off it goes toward the inner Solar System. Only when the dark, frozen chunk gets close enough to the Sun do we see the phantasmal glow of a comet.

Strangely enough, to the best of our knowledge, your crazy uncle's picture of cometary origins is probably correct. The Oort Cloud—sometimes called the Öpik-Oort Cloud—is what we call this distant hypothetical population of icy bodies. It contains potentially trillions

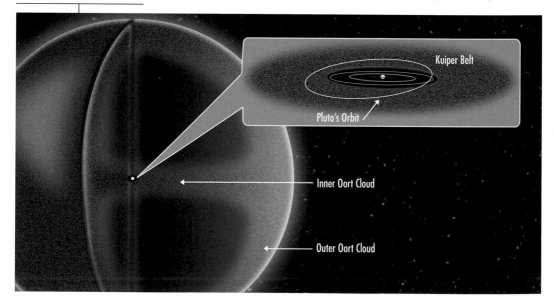

Kuiper Belt

Pluto's Orbit

Inner Oort Cloud

Outer Oort Cloud

of cometary nuclei—the icy insides of comets. The combined mass of these small distant chunks of ice may be more than 40 times the mass of the entire Earth. Although they are small, there certainly are a lot of them.

The idea of a reservoir of comets in a distant region of the Solar System was first suggested in 1932 by Estonian astronomer Ernst Öpik. Later, Dutch astronomer Jan Oort developed a similar idea to explain his observations of long-period comets—comets that take more than 200 years to make one orbit around the Sun.

Oort noticed several curious things about the long-period comets. First, none has an orbit that would lead us to believe it came from outside our Solar System—they are all in orbit about our Sun. Second, their orbital characteristics suggest that most first-time comets originally had roughly the same maximum distance from the Sun (called aphelion)—they start out from approximately the same distance before being perturbed into the inner Solar System. Finally, long-period comets appear to be rushing in toward the Sun from all directions equally—they are likely randomly distributed in a spherical cloud. These observations led Oort to suggest a vast and distant source region for the long-period comets.

Comet Hale-Bopp, seen here in the Croatian sky while approaching aphelion in 1997, is a long period comet typical of those thought to come from the Oort Cloud.

The Oort Cloud isn't the only potential source of small icy bodies in our Solar System. In fact, the outer Solar System is a complex region with different reservoirs of icy bodies. The innermost reservoir is the Kuiper Belt, whose best-known member is the recently demoted dwarf planet Pluto. The Kuiper Belt extends from Neptune's orbit at 30 AU from the Sun out to a distance of roughly 55 AU. The orbits of Kuiper Belt objects (KBOs) have been found to be stable, however, so the Kuiper Belt is an unlikely source for long-period comets.

Overlapping the Kuiper Belt is the Scattered Disk, whose members come as close as 30–35 AU to the Sun but whose highly eccentric (noncircular) orbits can extend out farther than 100 AU. Eris, the largest dwarf planet found so far, is an example of a Scattered Disk object. Scattered Disk objects (SDOs) have unstable orbits that can be influenced by Neptune's gravitational attraction. Because of this, the Scattered Disk is thought to be the most likely source region for the majority of comets whose orbital periods are less than 200 years (short-period comets). Orbits of SDOs can't account for the presence of the long-period comets, however.

After more than three-quarters of a century of study and despite what we know of long-period comets, the existence of the Oort Cloud remains only a working hypothesis. Why? Because we haven't actually seen it. The Oort Cloud's icy members are extremely distant, generally thought to be somewhere between 2,000 and 200,000 AU from the Sun. So although astronomers have observed objects in the much nearer Kuiper Belt and Scattered Disk, no definitive Oort Cloud object has yet been identified.

With potentially trillions of bodies available, it's only a matter of time before we're able to confirm the existence of the Oort Cloud through observation. Advances in telescope technology and more complete sky surveys are leading to the discovery of increasingly more distant objects at a record pace. One potential Oort Cloud candidate is the detached object known as Sedna.

Our Sun is merely a distant bright star in this artist's concept of the detached object Sedna. If you could hold a pin at arm's length while standing on the surface of Sedna, the head of the pin would block out the disk of the Sun. The planetoid's very distant and highly elliptical orbit makes it a potential candidate for classification as an inner Oort Cloud object.

Discovered in 2003, Sedna has a highly eccentric orbit that takes it nearly 1,000 AU away from the Sun. Even at the point in its orbit closest to the Sun (called perihelion), Sedna is still more than 75 AU away, well outside the boundary of the Kuiper Belt. In fact, Sedna is at all times so distant that Neptune's gravitational influence on the icy body is insignificant—Sedna is considered "detached" from the main Solar System. For this reason, the discoverers of Sedna argue that it should be considered not a part of the Scattered Disk but instead a member of the (much nearer than previously thought) inner Oort Cloud.

How did a reservoir of protocomets such as the Oort Cloud form, and why do its members occasionally come hurtling through the Solar System? Ironically, it is thought that these distant icy bodies were initially formed much closer to the Sun. After condensing out

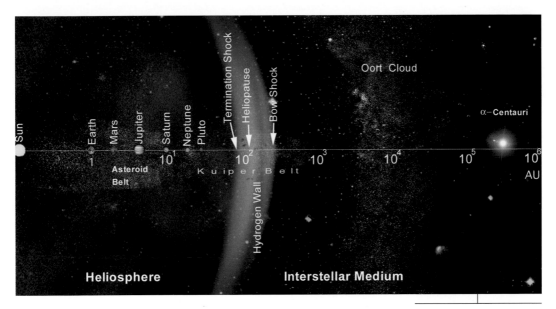

Distances from our Sun extending out to 1 million AU. The scale is a logarithmic one, where each distance mark is 10 times farther out than the previous. The outer edge of the Oort Cloud marks the limit of our Sun's predominant gravitational influence. Nearby star systems like α-Centauri, our closest stellar neighbor, may perturb Oort Cloud bodies and send them into the inner Solar System.

of the same solar nebula that formed the planets, these small icy planetesimals were flung out into the far reaches of the Solar System by gravitational interactions with the freshly formed giant planets. Those bodies that were not ejected from the Solar System outright or weren't lost to subsequent collisions eventually found themselves in quasi-stable orbits loosely bound to the Sun.

The Sun's hold on these protocomets is so tenuous that they are susceptible to very small perturbations. A distant passing star or massive molecular cloud, or even galactic tides—due to variations in our galaxy's (the Milky Way's) gravitational field—can dislodge an icy chunk and send it careening toward the Sun to become a long-period comet.

Precisely how many bodies inhabit the Oort Cloud, we don't know. For now, we can see only the tiny fraction of Oort bodies that graze the inner Solar System. As the late-20th-century astronomer Carl Sagan might say, there may be "billions and billions" of them out there. In what seems like a giant snowball fight—icy bodies tossed from the inner to the outer Solar System and then back again—the hypothetical Oort Cloud contains the largest collection of bodies in the Solar System.

A composite Hubble Space Telescope image of Jupiter and multifragment Comet Shoemaker-Levy 9. The dark spot on Jupiter is the shadow of Jupiter's moon Io, located to the right of the shadow in this image. Apparent sizes of Jupiter and the comet have been modified for illustration purposes.

It was a questionable night for observing with a telescope. Thin cirrus clouds covered the sky as a storm approached on March 24, 1993. At Palomar Observatory in southern California, astronomers Carolyn Shoemaker, Gene Shoemaker, and David Levy patiently waited for the sky to clear. They were hunting for yet undiscovered comets and asteroids, a meticulous process that involved taking two long-exposure photographs of the same part of the night sky, each exposure separated by about one hour. If an object moved relative to the background stars, it would pop out when comparing the two photographs in stereo, much like a 3-D movie. The astronomers decided to use some left-over, partially exposed photographic plates and to save good plates for a clear night.

Their perseverance was rewarded. A weird smudge, not caused by the imperfect plates, appeared on two plates with Jupiter in the field of view. They had photographed a unique comet, named Shoemaker-Levy 9 (SL9), the ninth comet discovery by the observing team.

This comet was different in many respects. First, SL9 orbited the planet Jupiter rather than the Sun like most comets do. According to calculations of past trajectories, SL9 originally orbited the Sun in a highly elliptical trajectory. But in 1929 (the year Carolyn Shoemaker was born), SL9 passed too close to the giant planet and was captured by Jupiter's strong gravitational field.

Second, unlike the single nucleus of most comets, SL9 consisted of 21 distinct fragments in a line, giving it a squashed or "string of pearls" appearance. Jupiter's strong gravitational field was again the culprit. Inside a planet's Roche limit (roughly 2.4 times the planet's radius), orbiting bodies cannot withstand the planet's strong gravi-

tational force and will be ripped apart. SL9 passed within the Roche limit of Jupiter in 1992, and SL9's single cometary nucleus (~1–4 km in diameter) split into 21 smaller fragments.

Perhaps most startling, calculations showed that SL9 would collide with Jupiter in July 1994. It would be the first direct human observation of a cometary impact on another planetary body.

Excitement over the impending collision mounted. Although the impacts would occur just behind the limb of Jupiter from Earth's view, the giant planet's rapid rotation would cause the impact sites to become directly visible just minutes after impact. Astronomers around the world turned their telescopes to the giant planet. Major ground-based observatories coordinated their activities, and SL9 observations from Hubble Space Telescope and the Galileo spacecraft received top priority. Located closer to Jupiter and with a different line of sight, the Galileo spacecraft would be the only instrument directly to observe the impacts.

SL9's fiery end lasted six extraordinary days. Each of the 21 impacts exhibited three phases: a glowing meteor as the fragment

Hubble Space Telescope image of Comet Shoemaker-Levy 9 on May 17, 1994, two months before colliding with Jupiter. Fragment A leads the other fragments in orbit. The train of cometary fragments extends over 1.1 million km (710,000 miles).

Fireball of fragment G impact on Jupiter (left) detected by infrared telescope at Siding Spring Observatory, Australia. Sequence of fragment W fireball (right) observed by the Galileo spacecraft over a seven-second period.

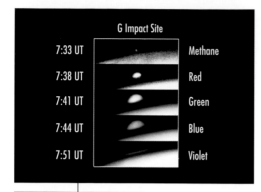

G Impact Site

7:33 UT	Methane
7:38 UT	Red
7:41 UT	Green
7:44 UT	Blue
7:51 UT	Violet

Hubble Space Telescope time sequence of fireball-to-splash phases of the fragment G impact. The brilliant plume was easily detected at near-infrared (methane) and visible wavelengths, and the splash phase can be seen in the last (violet) image. Jupiter's shadow produces an apparent gap between the fireball and the planet in all images.

shot through the Jovian atmosphere, an eruptive fireball as the fragment exploded in a hot plume of gas, and a stormy splash as plume material rained back onto the planet. On July 16, 1994, fragment A pummeled the giant planet at Mach 50 (50 times the speed of sound). The resulting explosion created a scorching fireball about half the size of Earth and four times hotter than the Sun's surface.

The largest impact occurred with fragment G on July 18. It scarred the planet with the most prominent and visible dark feature ever recorded on Jupiter. Large concentric rings produced by propagating waves and falling debris extended over 6,000 km (3,700 miles) from the central impact site. In total, the SL9 impacts yielded 10^{13} tons of TNT energy, equivalent to 700 million Hiroshima atomic bombs.

What if SL9 had hit Earth? A comet with a diameter of 1–4 km would cause global devastation. Life within a 5,000-km radius (an area larger than Asia) would be blasted by radiation from the fireball and ballistic reentry of ejecta. The ozone layer would be destroyed

Man on the Moon

It was Gene Shoemaker's dream to go to the Moon. Although a medical condition kept him from becoming an astronaut, Gene helped train Apollo astronauts on geology and cratering. During his remarkable career, Gene discovered over 800 asteroids and 20 comets, and he is credited with inventing the field of planetary geology. Gene died unexpectedly in 1997 in an automobile accident while crater hunting in the Australian outback.

Two years later, Gene finally made it to the Moon. A vial of his ashes piggybacked on the Lunar Prospector spacecraft, whose mission was to slam into the Moon at 6,000 km/hr (3,800 mph) and vaporize any water ice buried beneath the surface. Gene is the first human to be buried on a celestial body other than Earth. What a fitting tribute for the world's leading scientist on impact cratering to make his own crater on the Moon.

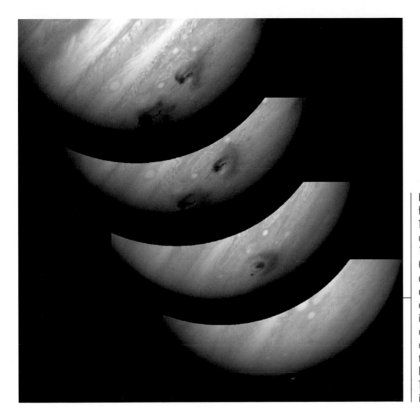

Hubble Space Telescope images of fragments G and L impact sites. Time progresses from lower right to upper left: 5 minutes after G impact; 1.5 hours after G impact; 3 days after G impact (1.3 days after L impact); and 5 days after G impact (3.3 days after L impact). The G impact plume extends beyond the limb in the earliest image. In the second image, the outermost ring of the G impact site is about the size of Earth. In the final two images, the G impact site is on the left and the L impact is on the right. Turbulent winds disrupt the circular impact sites over time.

for years. A thick dust cloud would envelop the entire Earth, blocking out sunlight for months, stopping photosynthesis, and causing worldwide starvation.

Fortunately, Earth's gravitational pull is smaller than Jupiter's, making a direct hit by a comet less likely—an SL9-sized comet will possibly hit Earth only every 0.2–2 million years. Even on Jupiter, a 1.5-km-diameter comet will strike the giant planet only about every 100 years. SL9-type impacts are rare events indeed.

Yet dedicated astronomers will still scan the night skies, often in less-than-ideal conditions, in search of the next great cometary impact.

In 1800, the Hungarian-born astronomer Baron Franz Xaver von Zach enlisted the aid of 23 of his fellow astronomers from around the world. His goal? To find a hitherto unseen planet between the orbits of Mars and Jupiter. He was looking in the part of the Solar System between those two planets based on the predictions of a highly effective planet-finding rule. The rule matched the positions of *all* of the known planets, so everyone *knew* that a planet must be there. By organizing one of the first systematic, large-scale, coordinated searches of the night sky, he hoped to go down in the annals of history as the chief discoverer of the new planet.

Known today as the Titius-Bode rule, the "miraculous" planet-finder relationship was based solely on the geometric observation that a planet's distance from the Sun is roughly double that of the one before it. Application of the rule to the planets at the time matched the distances to the four inner terrestrial planets (Mercury, Venus, Earth, Mars) fairly well. However, the large gap between Mars and

In this artist's depiction, a large spherical planetoid—much like the dwarf planet Ceres—can be seen in the main asteroid belt, a zone between the orbits of Mars and Jupiter filled with rocks and dusty debris in orbit about the Sun.

Jupiter could only be accounted for by "skipping a planet" before continuing the rule for Jupiter and Saturn (Saturn was the farthest known planet during the time that the rule's planet-finding prowess was being championed).

William Herschel's 1781 discovery of the planet Uranus lent further credence to the Titius-Bode rule. The distance of the newly discovered planet matched the rule's prediction almost exactly. The apparent validity of the rule led astronomers, and especially German astronomer Johann Bode (one of the namesakes of the planet-finding rule), to conclude that there must be an undiscovered planet between the orbits of Mars and Jupiter.

A Titius-Bode Scorecard		
Name	Predicted	Actual
Mercury	0.4	0.39
Venus	0.7	0.72
Earth	1.0	1.00
Mars	1.6	1.52
Ceres	2.8	2.77
Jupiter	5.2	5.20
Saturn	10.0	9.54
Uranus	19.6	19.19
Neptune	38.8	30.07
Pluto	77.2	39.48

The Titius-Bode rule predicts the distances of planets from the Sun, given here in astronomical units (AU). While Ceres (dwarf planet and largest of the asteroids) is fit by the rule, predictions fail to match ice giant Neptune or the dwarf planet Pluto.

So with the planet-finding rule in hand, the hunt for the missing planet was on! Ironically, it wasn't one of von Zach's *Himmelspolitzei* (celestial police) who discovered the "missing planet." In 1801, lone observer Giuseppe Piazzi, director of the Palermo Observatory, spotted what he first thought to be a comet (but hoped "might be something better") at the distance predicted for the missing planet. This small body was given the name Ceres, and it was the first of what eventually came to be known as asteroids.

We now know that there are hundreds of thousands if not millions of small bodies in orbit about the Sun in a zone between the orbits of Mars and Jupiter. This zone has been called the asteroid belt since at least 1850, even though only 13 asteroids were known at the time.

As more bodies have been discovered, the validity of the Titius-Bode rule has lost its luster: the planet-finding predictions failed to match the 1846 discovery of the planet Neptune . . . and all subsequent bodies. In fact, there is no strong scientific basis for why the Titius-Bode rule matches as well as it does for the planets it *does* match. Some astronomers dismiss the rule as mere coincidence, while others see the possibility of orbital interactions between planets leading to some sort of regular spacing during planet formation.

When it became clear that there were several minor planets instead

The main-belt asteroids lie between the orbits of Mars and Jupiter. The two largest asteroids, Ceres and Vesta (not specifically shown here), together account for over 40% of the total mass of the asteroid belt. Orbital distances are not to scale.

of one "missing planet," it was suggested that the small bodies were the remains of a larger planet that was broken up in some early, cataclysmic event. Although initially accepted with little question, the broken planet hypothesis has not withstood the test of time. Today, the asteroid belt is thought to be left over from the formation of the Solar System: planetesimals unable to coalesce further into larger bodies.

Gravitational interaction with nearby, massive Jupiter is the primary reason the asteroid belt has stayed chunky. Jupiter's strong perturbations prevent formation of a planet. Either planetesimals are sped up to the point that collisions result in them breaking apart rather than sticking together, or they are boosted into unstable orbits and ejected from the asteroid belt altogether. The Kirkwood gaps are evidence of such gravitational house cleaning. The present-day asteroid belt probably contains only a small fraction of the primordial belt.

Despite how the asteroid belt is often depicted (including the opening image for this chapter), there's not much "stuff" actually in the belt. In terms of mass, there is less than 5% of the amount of material that makes up Earth's Moon. While there are likely millions of individual small bodies, they are spread out over such a wide volume that there is quite a bit of empty space between them. Instead of dodging myriad swirling chunks, as in the movies or in video games, you'd need to try very hard to hit (or get hit by) an asteroid. In spite of initial concerns for early missions, none of the spacecraft to traverse the asteroid belt, so far, has experienced any difficulty. For NASA's Galileo mission, special efforts had to be made to find an asteroid near enough to fly by on the way to Jupiter.

Not all asteroids are confined to the asteroid belt, and not everything in the asteroid belt is an asteroid! At the outer edge of the asteroid belt, astronomers have spotted a few examples of what appear to be comets. Unlike regular comets, the main-belt comets have very circular orbits within the asteroid belt and exhibit a dust tail for only part of their orbit—around perihelion, the closest approach to the Sun.

Meanwhile, small rocky bodies have been spotted in several

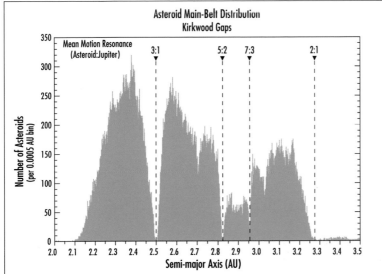

When you count the number of main-belt asteroids at given distances from the Sun, certain distances have many fewer asteroids. These gaps correspond to orbital resonances with Jupiter and have been cleared out by the giant planet's gravitational perturbations.

Asteroids visited by spacecraft so far, shown to scale. The name, size, visiting spacecraft, and year visited are given under the image of each asteroid. NASA's Dawn mission will orbit asteroids 4 Vesta and 1 Ceres in 2011 and 2015, while ESA's Rosetta spacecraft will fly by 21 Lutetia in 2010.

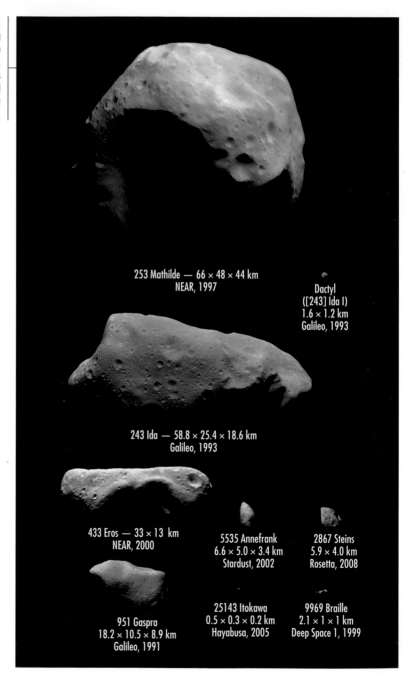

253 Mathilde — 66 × 48 × 44 km
NEAR, 1997

Dactyl
([243] Ida I)
1.6 × 1.2 km
Galileo, 1993

243 Ida — 58.8 × 25.4 × 18.6 km
Galileo, 1993

433 Eros — 33 × 13 km
NEAR, 2000

5535 Annefrank
6.6 × 5.0 × 3.4 km
Stardust, 2002

2867 Steins
5.9 × 4.0 km
Rosetta, 2008

951 Gaspra
18.2 × 10.5 × 8.9 km
Galileo, 1991

25143 Itokawa
0.5 × 0.3 × 0.2 km
Hayabusa, 2005

9969 Braille
2.1 × 1 × 1 km
Deep Space 1, 1999

regions of the Solar System. There are Trojan asteroids, which spend their time hanging out near Jupiter's Lagrange points (locations where the gravitational forces of the Sun and Jupiter exactly balance the centrifugal force of an asteroid's orbit—there are only five such points); a few Near Earth Asteroids (NEAs), which can pose some impact danger to Earth; unstable Centaurs, which orbit the Sun at distances between Jupiter and Neptune; and millions, if not billions, of rocky/icy bodies found in the Kuiper Belt, Scattered Disk, and Oort Cloud.

Today, the asteroid belt holds promise as a link back to the early Solar System. The remaining asteroids in the belt are likely leftover planetesimals, the building blocks of the inner planets. Vesta, the second most massive asteroid in the belt, is dry and "evolved" (old looking), while the dwarf planet Ceres shows evidence of water and perhaps a thin atmosphere. By visiting asteroids of different sizes and compositions, NASA's Dawn mission and ESA's Rosetta spacecraft will provide new clues to the evolution of our Solar System.

From missing planet to broken planet to planet that never formed, our understanding of the asteroid belt evolves as continued observations and exploration provide new data. All because a simple geometric relation, which probably has no scientific validity, led astronomers to look for a planet that wasn't.

Earth-Shattering Impact—Birth of Our Moon?

The Task . . . Your mission, should you choose to accept it, will be to solve one of the toughest riddles in the history of the Solar System. It won't be easy—the clues are scarce—and the final conclusion could be Earth shattering. You must uncover the origin of our Moon.

The Implications . . . This is no small matter. In fact, the Moon is quite big. The other terrestrial planets have nothing that really compares—Mars has two tiny asteroid-like satellites and Mercury and Venus lack moons altogether. Except for Pluto's moon Charon, our Moon's mass relative to the size of its planet is greater than that of any other satellite in the Solar System. Yet the Earth/Moon system

Most of the scarring on the Moon's battered surface occurred during the Late Heavy Bombardment era, roughly 3.8–4.1 billion years ago, when planetesimals slammed into the young Moon to create numerous craters. The dark areas are maria (seas), large impact basins filled in by dark basaltic lava after impact. Brighter regions are ancient highlands littered with smaller craters. Ironically, a large impact may have created the Moon in the first place.

is quite distinct from the dwarf planet system; our Moon alone is more massive than Pluto and Charon combined!

The gravitational force on Earth by the Moon—you see its influence daily in the ocean tides—also causes a remarkable climate effect. Earth has a stable obliquity of 23.4°, which produces our highly predictable seasons. Over a period of 41,000 years, this tilt rocks back and forth ever so slightly by ±1°. Without the persistent correctional tug of a large moon, the wobble would be much more pronounced—Mars's tilt, for example, may vary from 15° to 80°! Earth's stable axis keeps our climate relatively constant over long periods, allowing liquid water to flow and complex life to flourish.

In other words, the birth of a large Moon changed the evolution of our home planet. You might not be here without the Moon.

The Clues ... Your first clue to the origin of our Moon involves its interior composition. Unlike the terrestrial planets, the Moon is relatively uniform. There is a significant absence of heavy stuff— the Moon's density is closer to that of the Earth's mantle than to the Earth as a whole. Lunar seismic data from the Apollo missions, gravity and magnetic field measurements by orbiting spacecraft, and ongoing lunar laser ranging experiments (yes, scientists shoot lasers at the Moon) all suggest an unusually small iron core that makes up less than 3% of the Moon's mass.

The next clue comes from lunar rock samples and lunar meteorites. Surprisingly, Moon rocks have oxygen isotopes identical to those found in the Earth's mantle. Because they are enriched in elements with high melting temperatures (like uranium and titanium) and depleted in compounds that easily vaporize (like water and carbon dioxide), lunar rocks likely solidified from a hot magma ocean on the Moon's surface.

Finally, any theory of the Moon's origin must account for the total angular momentum of the Earth/Moon system. This angular motion includes the rotation of the Earth (an Earth day), the rotation of

As the Galileo spacecraft flew by Earth in 1992, it took this remarkable photo of the Earth and Moon. The Moon (upper left) is moving from left to right. The large size of the Moon relative to Earth helps stabilize Earth's axial tilt and thus our climate.

Astronaut Buzz Aldrin sets up the Passive Seismic Experiment Package during the Apollo 11 mission in 1969 (left). Lunar seismic measurements provided important information about the Moon's interior. This lunar rock sample (right) was collected from a highland region during the Apollo 16 mission in 1972. It is a classic example of a lunar breccia, a conglomerate rock formed when meteorites break apart the surface and weld it back together under the intense heat and pressure of impact. The white fragments are feldspar minerals 4.5 billion years old that formed when the Moon's crust first solidified.

the Moon (a lunar day), and the mutual rotation of the Earth/Moon system about its center of mass (a lunar month). Today, the same side of the Moon always faces Earth—its rotation is tidally locked to its orbit—so a lunar day equals a lunar month. But this hasn't always been the case.

The Hypotheses . . . You should consider at least the following four hypotheses while seeking the origin of our Moon. None of them has been proven. Feel free to develop alternative explanations that fit the available facts.

First suggested in 1878 by George Darwin (son of biologist Charles Darwin), the *fission hypothesis* states that the Moon popped out of the Earth due to high-speed rotation of the planet. A few years later, Osmond Fisher suggested that the Pacific Ocean basin was the remnant of this birth. This hypothesis can explain the Moon's similarity to the Earth's mantle, but the angular momentum required to spin off a chunk of material is far too great. Plus, we now understand that the vast Pacific basin results from plate tectonics and continental drift.

The *capture hypothesis* claims that the Moon formed in a different part of the Solar System but was gravitationally captured as it strayed too close to Earth. Unfortunately, the mantle-like composition of the Moon cannot be explained by such a capture, and conservation of angular momentum causes problems for the hypothesis. Like dance

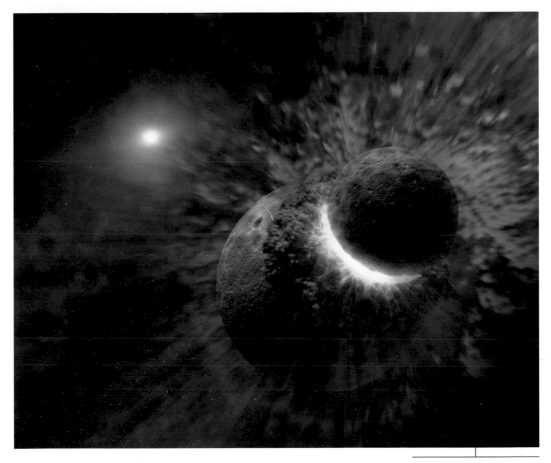

The giant impact hypothesis states that early in the Solar System's history, a Mars-sized body possibly smashed into the young Earth, liquefying the Earth's surface and splashing hot debris into orbit. The Moon formed from a fraction of this material; the rest rained back onto Earth.

partners rapidly approaching each other from different parts of the floor, the Earth and Moon would have to slow down unreasonably quickly to combine as a single dance team with the correct amount of spin.

The *double planet hypothesis* purports that the Earth and Moon formed together as a binary system at the same distance from the Sun. This hypothesis suffers from the iron core problem—why should the Earth have so much iron and the Moon so little, if they formed in the same way? Likewise, the double planet hypothesis suggests that Earth and Moon should have similar proportions of volatiles (easily vaporized material).

The *giant impact hypothesis* argues that a Mars-sized object slammed into the early Earth, obliterating the Earth's crust and flinging mantle material into space. The Moon coalesced from a hot brew of this splashed-out, iron-depleted debris. To get the correct angular momentum, the preimpact Earth must have been spinning retrograde, or backward, like Venus does today. This hypothesis readily explains the uniformity of the Moon, the mantlelike composition of lunar rocks, the hot origin of the Moon, and the current angular momentum of the Earth/Moon system.

Yes, it's popular, but the giant impact hypothesis has its problems. Although a high-energy impact can explain the release of volatiles, the ratios of volatiles remaining on the Moon don't match current theoretical predictions. Furthermore, the iron-depleted Moon actually has *more* iron than the Earth's mantle. Did the extra iron come from the Mars-sized impactor, or did some of the Earth's core also get splashed out?

The Secrets . . . As you can see, the solution will not be simple. We have very little information to go on. Any residual scar on Earth from the giant impact has long been erased—the impact would have melted the Earth's surface and reset its geologic clock. Indeed, the oldest known Earth rock is about 100 million years *younger* than the oldest Moon rock. Earth is going to provide very little help.

The Moon also has her secrets. We have collected lunar samples from only a handful of locations. Chemical analysis of samples from a different site might yield rocks with higher iron content, invalidating the giant impact hypothesis. More advanced seismological measurements will give us a better understanding of the structure of the Moon's interior, while high-precision gravity and magnetic field measurements from lunar orbit can provide meaningful clues about the Moon's core.

In essence, you must return to the Moon to collect the evidence needed to solve this mystery. Good luck, and keep us informed of your progress. The human race depended on it.

Part Six

Electro Magneto Extremo

Supertwisted Magnet—Our Sun

The greatest disaster to strike human civilization in centuries might be just around the corner. It would come as an intense blast of X-rays from outer space. Classified military satellites—zapped. Electrical power grids—toast. Sewage and water treatment—stagnant. Air traffic control—silent. Our high-tech society would be crippled in an instant, potentially costing the world economy over $20 trillion. Even worse, the intense X-ray radiation would annihilate up to 10% of the protective ozone layer, and incidents of skin cancer would rise to historic proportions.

No, it's not the start of an alien invasion. In a twisted series of events, this modern-day catastrophe would be incited by the most extreme magnet in our Solar System—the Sun.

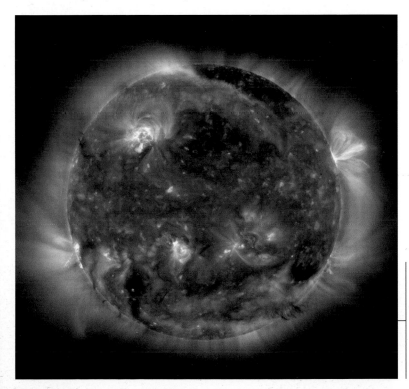

The most extreme magnet in the Solar System. This composite image at three different wavelengths of extreme ultraviolet radiation (blue = 17.1 nm, yellow = 19.5 nm, red = 28.4 nm) shows a magnetically active Sun: bright solar flares, magnetic arcs, and hot plasma streamers extending into the corona.

Two sunspots collided on December 13, 2006, producing major solar activity. The top ultraviolet image shows a bright flare composed of ionized hydrogen (protons). The middle image at visible wavelengths captures the colliding sunspots with interacting filaments. The bottom image shows the line-of-sight magnetic field, where white indicates positive polarity and black is negative. A contorted magnetic field emerges near the collision. Six Earths would fit inside the larger sunspot.

The Sun actually contains millions of magnets, each one produced by turbulent bubbles of plasma the size of Texas. As electrically charged particles—the plasma—circulate in the roiling outer layers of the Sun, they generate local magnetic fields. Sometimes you can find simple dipole magnets that behave like a bar magnet, but more often you find complex arrangements of dipoles, quadrupoles, and more.

These strong magnetic fields often emerge from the Sun's fiery surface, or photosphere, within relatively dark regions called sunspots. Plasma within the spots gets trapped by the intense magnetic fields and cannot participate in the boiling convection of the Sun. The trapped plasma cools rapidly and produces less light (and thus appears darker) than the surrounding hot convecting gases.

Since charged particles flow along magnetic field lines, glowing plasma often outlines the magnetic field structure. Bright magnetic filaments appear along the outer edge of sunspots. From above, the filaments look like radiant eyelashes surrounding a dark central eye. From the side, the plasma traces beautiful magnetic arcs streaming out from the sunspot.

But these delicate, peaceful-looking arcs are deceptive. Magnetic field lines change with the moving plasma, twisting and morphing into new configurations.

Imagine a rubber band (or better yet, go get one) stretched across the thumbs and index fingers of each hand to create a loop. Flip one hand so that your thumb and index finger change places. The twisted rubber band now has two loops with an intersection in the middle. In a magnetic field, the two loops violently separate from each other at the connection point—a magnetic short circuit—and release tremendous amounts of energy in the process.

When this separation and magnetic reconnection occur on the

Sun, look out! Near sunspots with highly twisted magnetic fields, bright solar flares explode with energies equal to 10 million atomic bombs. Ionized particles accelerate to almost the speed of light, and temperatures sizzle at over a million degrees! The most violent solar flares—called X class—release enough X-ray radiation to trigger worldwide radio blackouts on Earth.

Magnetic reconnection also blasts charged particles off the Sun to produce enormous coronal mass ejections (CMEs). Often associated with solar flares, CMEs hurl as much as 10% of the Sun's corona (outer atmosphere) through space at breakneck speeds. The normal solar wind takes three to four days to reach Earth, but the fastest CMEs hit Earth in only 17 hours! These intense plasma balls squeeze Earth's magnetosphere to produce strong geomagnetic storms with dazzling auroras.

On November 4, 2003, a record-setting solar flare maxed out satellite sensors with a rating of X28+ (at least 28 times more powerful than energetic X1 storms). Luckily, it wasn't aimed directly at Earth. Such was not the case in 1859, when X-rays and high-energy CME protons from the first observed solar flare made a direct hit. Earth's magnetic field usually buffers us from this kind of onslaught, but the magnetic field of the 1859 solar superstorm was aligned in the *opposite* direction from Earth's magnetic field. This allowed easy north-south connection of the two magnetic fields, and as you know, extreme things happen when magnetic fields connect. Telegraph offices—high-tech features of the day—caught fire as wires became overloaded with electric current induced by the huge magnetic pulse. A similar event today would wreak havoc on our electricity-dependent world.

On average, the Sun undergoes an 11-year cycle in sunspot activity. The most recent cycle began with a sunspot minimum (and magnetic polarity reversal) in January 2008. This solar cycle has

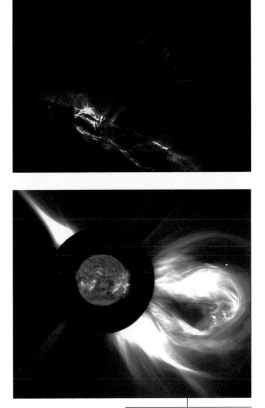

Hot plasma flows along curved magnetic field lines emanating from the Sun's surface. The bright arc is a modest M-class flare, 10 times less intense than extreme X-class flares.

A huge coronal mass ejection (CME) sends charged particles into space. To observe CMEs, a red occulting disk is used to cover the Sun. An ultraviolet image of the Sun taken from about the same time is superimposed on the red disk.

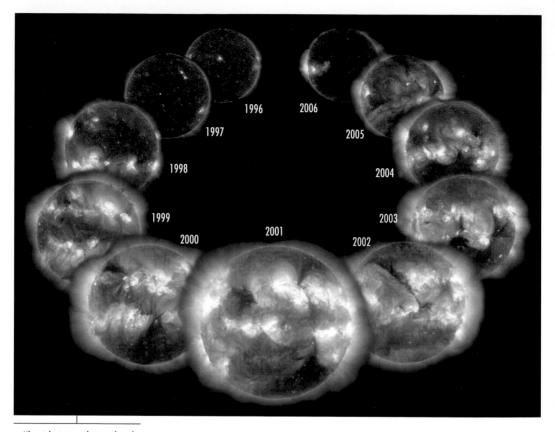

Ultraviolet images showing the solar cycle from solar minimum (1996) to solar maximum (2001) to near solar minimum again (2006).

been remarkably calm—the quietest in nearly 100 years—producing speculation that we could be facing extremely low sunspot activity similar to the Maunder Minimum observed between 1645 and 1715. This period coincided with the bitterly cold Little Ice Age in Europe and North America.

But it's more likely that the recent quiet period is a delayed onset of solar activity. New groups of sunspots are beginning to fire up at the Sun's equator. Solar flares and CMEs will be much more prevalent, and thus potentially more dangerous, as the cycle heads toward sunspot maximum. With a more active Sun, chances of a monster blast racing toward Earth with the "wrong" magnetic polarity increase dramatically. The next supertwisted solar max is expected in 2013.

Blowing a Bubble in Space—The Solar Wind

What if we told you that you are living in the Sun? No, not residing in a beach town in Southern California or along the Spanish coast, but actually living *inside* the Sun? At first you might think that we've gone bonkers. After all, the Sun is a good 150 million km (93 million miles) away. But what if we mentioned that there is a part of the Sun that you can't easily see, a big bubble of plasma that marks the Sun's electromagnetic territory in space? This bubble—the heliosphere— is formed by an extended, billowing portion of the Sun's atmosphere called the solar wind. You live inside the heliosphere, bathed in solar plasma along with all the other planets, asteroids, Kuiper Belt objects, and more.

In many respects, the solar wind is just what the name implies: a fiery gale blowing from the Sun. It starts in the Sun's atmosphere, the transparent, outermost region known as the corona. Although the full details of the physical processes aren't fully understood, observations show that the plasma of the Sun's corona is superheated—more than two orders of magnitude hotter than the photosphere below—and accelerated away from the Sun. Both hot and fast, the ionized solar wind particles race outward at temperatures in excess of 1.5 million degrees and speeds between 300 and 900 km/s (700–2,000 mph).

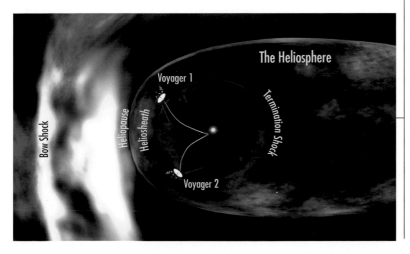

After completing their tours of the outer planets, NASA's Voyager 1 and 2 spacecraft continue their journey into the outer reaches of the Solar System. Their current mission: to explore the heliosphere. Voyager 1 crossed the termination shock into the heliosheath in 2004 at a distance of 94 AU from the Sun. Voyager 2 crossed the termination shock (or was crossed by it, as the shock moved back and forth several times) three years later at a distance of 84 AU. Although both continue toward the heliopause, it may take 10–20 years before either of these remarkable spacecraft reaches that boundary.

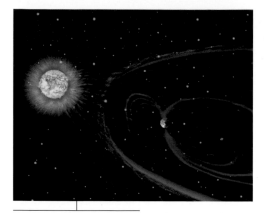

Earth's dipole magnetic field deflects and is distorted by the solar wind blowing outward from the Sun. Charged particles that make it through to the Earth's magnetic field can stream poleward to create auroras.

By the time the solar wind reaches Earth, it has cooled to a mere few hundred thousand degrees, yet it still howls toward us at an average speed of 400 km/s (900 mph).

On the other hand, as hot and fast as the solar wind is, there's not actually very much *stuff* in it. The density of Earth's atmosphere is many times greater than that of the solar wind. With only a few particles per cubic centimeter, the strength of the "breeze" as the solar wind reaches Earth wouldn't even ruffle your hair.

You shouldn't let the low density fool you, however. This gentle puff still packs quite a punch. The outward flow *is* able to blow the gases escaping from comets into long ion tails that always point away from the Sun. It distorts the magnetospheres of planets into long teardrop shapes and sparks brilliant auroras as ionized particles hit the atmosphere. The solar wind can strip away atmospheric constituents from planets both with and without

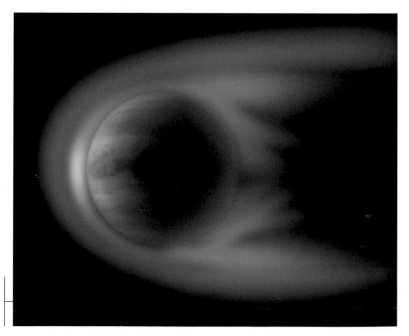

Solar wind plasma can strip material from the atmospheres of planets. This is one possible mechanism for how Venus lost its water.

magnetic fields. On airless bodies such as our Moon, the solar wind weathers and darkens their old, unprotected surfaces.

As a plasma, the solar wind is an excellent conductor and thus readily interacts with the magnetic field of the Solar System's largest magnet (the Sun). The Sun's magnetic field is "frozen" into the plasma and is carried outward by the solar wind. This not only creates the interplanetary magnetic field that permeates the Solar System but also causes the Sun to slow its rotation. Both by transferring angular momentum outward via the solar wind particles and by magnetic braking (tension in magnetic field lines coiled by the Sun's rotation), the Sun's spin has been significantly reduced over time.

But the effects of the solar wind don't stop there. Once far enough from the Sun, the outward-flowing solar wind travels supersonically until it is forced to slow down by the plasma of the interstellar medium (that's right, stellar winds from *other* stars). The transition from supersonic flow to subsonic flow results in the termination shock, a sudden change in density and temperature of the solar wind plasma. From there, the subsonic solar wind continues outward until it reaches the heliopause, the outer boundary of the Sun's bubble.

Beyond the heliopause, the flow of interstellar plasma and interstellar magnetic fields dominate over that of the solar wind. You might think that the Sun's influence would be finished. But on the leading edge of the Sun's path through outer space, the heliosphere compresses the interstellar plasma into a bow shock, much like a ship producing a bow wave as it cuts through the water. This compressed plasma glimmers with ultraviolet radiation called Fermi glow.

Our knowledge of this electromagnetic bubble isn't simply theoretical. In 2000, the Hubble Space Telescope detected the faint Fermi glow of the bow shock *outside* the heliosphere, perhaps as far as 230 AU from the Sun. More significantly, the termination shock *within* the heliosphere has been directly felt. Both of NASA's Voyager spacecraft have crossed the termination shock

Make a model heliosphere in your kitchen sink. The dry basin of the sink represents the interstellar medium. The water flowing outward after the falling stream hits the basin simulates the solar wind. The "hump" of water where flow changes from "rapidly outward" to "back toward the drain" models the termination shock. Where the water loses its battle with gravity and can no longer spread into the basin represents the heliopause.

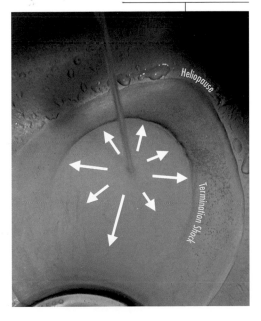

A ribbon of energetic neutral atoms (ENAs) discovered by the IBEX spacecraft appears to coincide with regions where the interstellar magnetic field lines are tangent to the heliopause. Red indicates the highest number of ENAs measured by the spacecraft. Yellow and green indicate lower numbers of ENAs, and blue and purple show the lowest number of ENAs. The heliosphere and the interstellar field lines are both deflected where they interact. Due to their trajectories, Voyagers 1 and 2 were unable to observe this ENA ribbon as the two spacecraft crossed the termination shock into the boundary region.

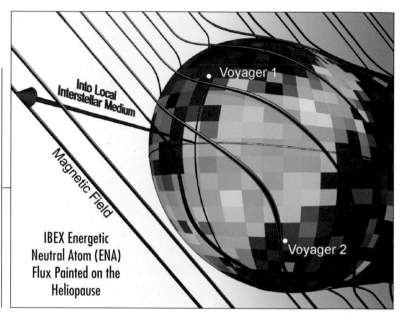

into the heliosheath. In another decade or two, Voyager 1 is destined to cross the heliopause into interstellar space to become Earth's first interstellar probe!

Fortunately, we won't have to wait for the Voyagers to reach the heliopause to learn more about the *extremely* big bubble being blown by our Sun. The Interstellar Boundary Explorer (IBEX) was launched in fall 2008 with the express purpose of mapping the heliopause and is already making fascinating new discoveries about the boundary of our Sun's bubble in space. Meanwhile, a suite of spacecraft including SOHO, WIND, STEREO A/B, and SDO continue to monitor the solar wind. After all, it's probably a good idea to keep an eye on things when you're living inside the Sun.

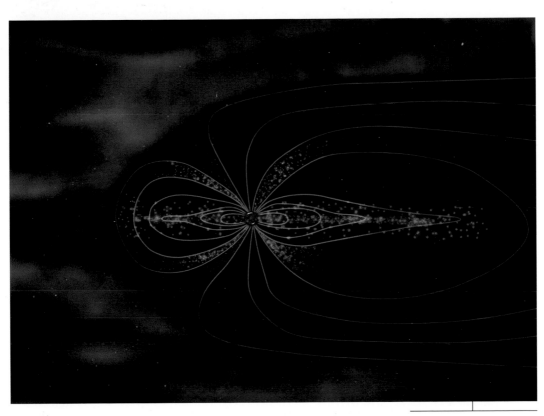

Have you ever stood amazed in front of a huge plasma-screen TV? Brilliant colors dancing across the display and details seemingly larger than life? Televisions seem to get bigger and brighter every day. The largest high-definition plasma TV in the world today can completely fill a living room wall. But the largest plasma screen in the Solar System—Jupiter's immense magnetosphere—dwarfs our television sets, and for that matter, everything else in the Solar System.

A plasma-screen TV basically works like this: Xenon and neon gases are trapped in tiny cells, called pixels, between two glass plates. Electrodes on the plates generate an electrical current through the cell and produce ionized gas—that's the plasma. Ultraviolet light is

Jupiter's magnetosphere separates plasma from the Sun (the solar wind) and plasma generated within the Jovian system. The volcanic moon Io populates the magnetosphere with particles (white dots) that become ionized (gold dots) by radiation and particle collisions. Jupiter's strong magnetic field (light blue lines) traps these charged particles, effectively producing a giant screen of plasma. The largest structure within the Solar System, Jupiter's magnetosphere extends well beyond Saturn's orbit.

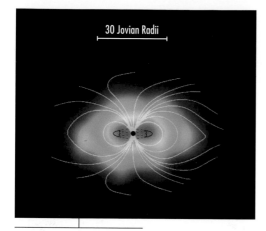

30 Jovian Radii

The Cassini spacecraft made Jupiter's invisible magnetosphere "visible" by capturing neutral atoms ejected from the magnetosphere by collisions with trapped ions. Charged ions travel along magnetic field lines (depicted in white), while neutral atoms are unaffected by magnetic fields and thus can escape the magnetosphere. Strong collisions occur near the Io plasma torus outlined in black. The Sun is to the left in this image.

emitted when plasma ions slam into one another. When this UV light hits the back glass plate, colorful things happen. Phosphors on the back plate absorb UV light and release different colors of visible light depending on the type of phosphor. This process happens a thousand times per second for each of the million or so pixels on a plasma TV.

Jupiter's extreme plasma "screen" is actually the boundary of Jupiter's magnetic influence, or magnetosphere. This mammoth magnetic bubble protects Jupiter from high-speed ions expelled from the Sun, causing the solar wind to flow around the planet in an elongated teardrop shape. Depending on the viewing angle (and if it could actually be seen with the unaided eye), Jupiter's magnetosphere would appear up to 15 times larger than the Sun to a backyard observer on Earth. It extends over 650 million km (404 million miles) downstream to Saturn's orbit and beyond, easily making Jupiter's magnetic barrier the largest thing within the Solar System.

Amazing things happen inside the magnetosphere. Electrons zip around at close to the speed of light. Plasma trapped within Jupiter's powerful magnetic field collides to produce intense electromagnetic radiation, and brilliant light shows flash across the sky. It's like watching a plasma TV on steroids.

Perhaps the most unique feature in Jupiter's magnetosphere is caused by its moons. Because their orbits are entirely within the magnetosphere, the largest satellites are a major source of trapped plasma. As the most volcanically active place in the Solar System, the moon Io ejects copious amounts of sulfur and oxygen to produce a doughnut-shaped torus of plasma along its orbital path. Recently, a gas torus was also detected in Europa's orbit by the Cassini spacecraft, possibly caused by impacts on the icy moon's surface from high-energy particles from Jupiter.

Charged ions must travel along magnetic field lines, but they don't just move in straight lines—instead, they spiral around the magnetic field lines in a twisted, looping trajectory. Magnetic field lines converge near Jupiter's north and south poles. This means that ions are packed into a much smaller space as they get closer to the planet. These ions collide with Jupiter's atmosphere and produce spectacular auroral displays that streak across the planet. Radiant electrical footprints of Io, Europa, and Ganymede have been observed in Jupiter's auroras.

So why is Jupiter's magnetosphere so big? There are three main influences: solar wind, planetary rotation, and magnetic field strength.

Jupiter's magnetic field constantly duels with the solar wind at the magnetopause, the edge of Jupiter's magnetosphere. By the time it reaches Jupiter's distant orbit, the solar wind has become rather diffuse and thus exerts a relatively low pressure on the magnetosphere. The force of the solar wind at Jupiter is over 25 times weaker than at Earth, allowing Jupiter's protective magnetic bubble to expand outward to greater distances.

The interior of Jupiter rotates once every 10 hours. (In fact, precise measurements of Jupiter's rotation rate are made possible by the magnetosphere. Charged particles, stuck in Jupiter's magnetic field as it rotates with the planet, emit radio signals with the same period as the planetary rotation.) The rapid rotation actually causes the magnetosphere to stretch out into a spinning disk, not unlike how hand-tossed pizza dough spreads out as it is spun in the air.

But ultimately, it is the incredibly strong magnetic field of Jupiter (second only to the Sun's in intensity) that produces the most extraordinary plasma screen in the Solar System. Deep within the planet, ionized hydrogen and helium continuously roil in an intense pressure cooker caused by Jupiter's immense mass. The convecting ion stew generates magnetism 19,000 times greater than on Earth.

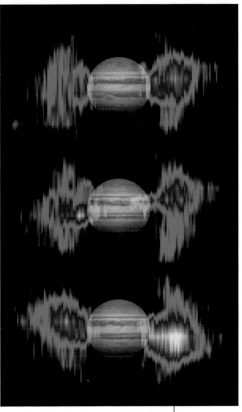

The Cassini spacecraft measured radio signals from high-energy electrons trapped within Jupiter's magnetosphere. The electrons race at 99% the speed of light and create intense radiation belts. The belts are tilted with respect to the equator since Jupiter's magnetic poles are not perfectly aligned with the geographic poles.

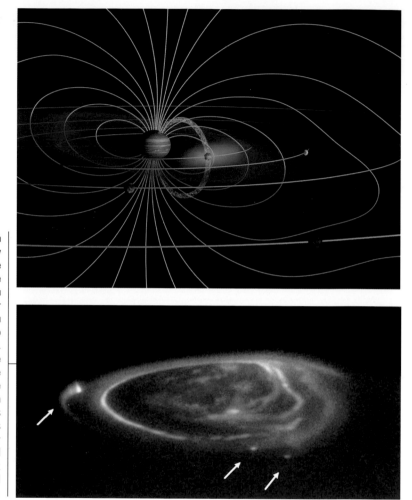

Inside the plasma screen: The Galilean satellites are electromagnetically coupled to the giant planet (top). The innermost moon Io leaves a dense torus of gas in its orbit, and Europa (the next moon out) exhibits a weaker torus. Magnetic field lines form a bridge for electric current (plasma) to flow from the large moons to Jupiter. The imprint of this magnetic bridge on Jupiter can be seen in this Hubble Space Telescope ultraviolet image of the planet's north polar region (bottom). The main aurora oval is centered on the magnetic pole. Arrows point to auroral footprints from Io (far left), Ganymede (lower center), and Europa (lower right center) that mark the locations of the magnetic field connecting Jupiter with the moons.

Without this strong magnetic field, Jupiter's magnetosphere would lack light-speed electrons, magnetic connections to its satellites, and brilliant auroral displays. Jupiter's plasma screen would be as bland as, well, a small black-and-white TV.

Most Radical Electric Light Shows—Auroras on Earth and Jupiter

Colors dance across the night sky—green, red, blue, violet—in dazzling rays and mystical arcs. Curtains of light shimmer and morph into enchanted shapes, only to fade inevitably into the background night. For the Inuit, dead souls of their ancestors journey across the sky with dramatic flourish. For the ancient Finnish, the mischievous firefox emblazoned the night sky with its fiery tail. For anyone who has witnessed an aurora in full glory, the experience is deeply magical.

Yet danger lurks during these intense light shows. Events that trigger this nighttime explosion of color can also disrupt radio communications, damage high-tech satellites, and potentially injure astronauts as they explore outer space.

Spectacular northern lights above Bear Lake, Alaska.

Aurora australis (southern lights) as observed from the Space Shuttle Discovery during the solar activity maximum in 1991. Vivid green and red colors of the aurora are produced by excited oxygen atoms 100–500 km (60–300 miles) above the surface.

Known as aurora borealis (northern lights) in the northern hemisphere and aurora australis (southern lights) down under, auroras occur in the upper atmosphere at high latitudes. They form brilliant ever-changing ovals that surround the magnetic poles. In "quiet" times, the aurora appears as a stationary diffuse glow, but during active times it can zip across the sky in the blink of an eye.

The key to producing an aurora is to get charged particles to hit the upper atmosphere. The basic sequence is this: a magnetic storm on the Sun injects high-energy electrons and protons into the Solar System. This solar wind of ionized particles carries with it the Sun's magnetic field, which interacts strongly with the Earth's magnetic field (and other planetary magnetic fields). Complex electromagnetic physics now happens: magnetic fields split and reconnect, charged particles accelerate, and vast amounts of energy are released. Here is where the threat appears—high-speed energetic particles can cripple the power grid or zap an unsuspecting astronaut. And the Earth's magnetic field directs these damaging particles in a spiraling high-speed trajectory toward the poles.

Luckily, the upper atmosphere shields us from this onslaught. When charged particles hit the upper atmosphere, they collide with neutral atoms and strip the atoms of their own electrons to produce

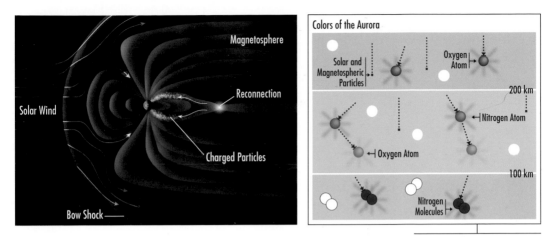

Colors of the Aurora

Magnetosphere

Reconnection

Solar Wind

Charged Particles

Bow Shock

Solar and Magnetospheric Particles

Oxygen Atom

200 km

Nitrogen Atom

Oxygen Atom

100 km

Nitrogen Molecules

The solar wind interacts with the Earth's magnetic field and wreaks electrical havoc. On the sunward side, charged solar wind particles can leak into the magnetosphere near the poles. Magnetic reconnection, often induced by enhanced solar activity, accelerates charged particles in the trailing lobe of the magnetosphere. Auroras occur when these charged particles slam into oxygen and nitrogen in the upper atmosphere to create ion radicals. Light is emitted when these ions recombine with free electrons.

ion radicals. A radical is an atom or molecule with an unpaired electron; it is highly reactive, desperately searching for another electron (atoms and molecules are chemically stable when their electrons exist in pairs). When the radical captures a free electron, it releases a photon of light. Oxygen atoms emit green and red light, while molecular nitrogen releases red, blue, and violet radiation—the mystic colors of the aurora!

We are just now learning the secrets of Earth's auroras. For instance, it has long been thought that aurora-producing charged particles come from a reservoir of trapped ions within the Earth's magnetosphere called the plasma sheet. However in 2007, the THEMIS suite of satellites observed for the first time a magnetic "rope" with Earth and solar magnetic fields intertwined. The rope channeled solar wind particles directly to the upper atmosphere and generated a spectacular aurora. Then in 2008, THEMIS discovered that strong auroras occur during magnetic "substorms" spawned by magnetic reconnection about one-third of the way to the Moon.

Although Earth's auroras are radical and electric, they probably are not the most radical electric light shows in the Solar System. This honor goes to the auroras with the most ion radicals—get it, the most radical—Jupiter's auroras.

Jupiter's auroras are, in a word, *extreme*. They produce enough energy to power all the cities on Earth. The auroras are so intense that they emit not only visible light but also radio, infrared, ultraviolet,

This X-ray image from NASA's orbiting Chandra Observatory shows high-energy auroras at Jupiter. Auroral emissions occur predominately from atomic and molecular hydrogen, although radiation from other molecular compounds has been detected. The auroras are larger than planet Earth.

and X-ray radiation. And the main auroral ovals are so large that they would swallow our own humble planet.

Like on Earth, Jupiter's auroras are produced by charged particles hitting the upper atmosphere, but the primary source of these charged particles is quite different. Rather than depending on the solar wind for high-speed ions, Jupiter begets its own ions through interactions with its moon Io. The continual stretching and tugging on Io by Jupiter's intense gravity causes the large moon to belch sulfur and oxygen gas. These gases become ionized, producing a doughnut-shaped torus of plasma in Io's orbit. Eventually, charged particles migrate throughout the entire magnetosphere of Jupiter.

Now for the radical part: Auroral emissions are caused by high-speed charged particles slamming into the atmosphere to create hydrogen radicals (Jupiter's atmosphere is composed primarily of

Ultraviolet images of Jupiter's auroras from the Hubble Space Telescope superimposed on a Hubble visible image of the entire planet. The main auroral ovals trace where magnetic field lines hit the upper atmosphere. The cometlike streaks in both auroras are footprints of Jupiter's moon Io.

hydrogen). But other radicals are also produced—emissions from ammonia, methane, ethane, acetylene, benzene, and radioactive hydrogen H_3 have been detected. It makes Earth's radicals of nitrogen and oxygen seem tame in comparison.

Because Io is such a major source of ions, the moon leaves its mark on Jupiter. The Voyager spacecraft detected strong electrical currents (flowing ions) of over 5 million amps connecting Io to Jupiter (bright sports-stadium lights draw only 9 amps of current). The result is a flying, supercharged auroral "footprint" on Jupiter with a fluorescent tail that stretches across the Jovian sky. The moons Europa and Ganymede also generate luminous auroral footprints that flash by at breakneck speeds over 16,000 km per hour (10,000 mph).

Now that's radical, dude!

Shocking Superbolts of Saturn

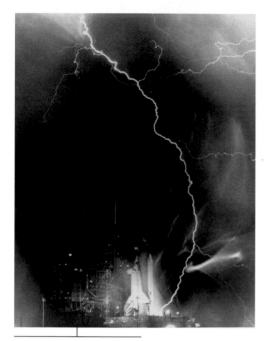

The Lightning Detection and Ranging (LDAR) system at Kennedy Space Center in Florida has measured as many as 425 lightning flashes per minute. The community of Lakeland, 160 km (100 miles) to the west, experiences the highest number of thunderstorms in North America. Space shuttle launches are often delayed due to extreme weather.

It's a crisp, clear winter's day. Your feet scuff across the carpet, picking up stray electrons along the way. As you reach for the doorknob, you feel a sharp pinprick on your finger and hear a crackle in the air. Roughly 2 amperes of current leave your finger as over a trillion electrons leap across the gap from you to the doorknob. It happens in a split second—you have just discharged 2,000–20,000 volts* of electricity in less than 100 nanoseconds—and your finger recoils in relatively slow motion. Shocked but unharmed, you blithely walk away from the electrical discharge.

Lightning strikes are an entirely different matter. Although faster than the blink of an eye, lightning bolts last over 300 times longer than the harmless doorknob discharge. With extremely high voltages (~100 million volts) and very large currents (~30,000 amperes), a single lightning bolt unleashes enough electrical energy to power the entire United States for a year. Temperatures reach a sizzling 28,000°C (50,000°F), almost five times hotter than the surface of the Sun. So it should come as no "shock" to you that lightning can be deadly; in 1939, a single stroke of lightning killed 835 sheep as they huddled under a tree. That's baaad!

Scientists don't exactly understand how thunderstorms become electrified, but the leading theory involves collisions of ice particles. Thunderstorms contain a mixture of water droplets, small ice crystals (snow), and larger ice particles (graupel and hail). Strong updrafts in the thunderstorm are able to keep graupel and hail aloft temporarily, but eventually the large ice particles fall toward the

---

*By comparison, a typical household electrical circuit draws a maximum of 20 amperes of current at 110–250 volts.

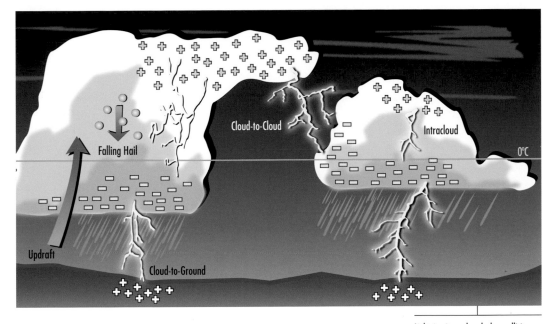

Falling Hail

Cloud-to-Cloud

Intracloud

0°C

Updraft

Cloud-to-Ground

Lightning is produced when collisions by ice particles cause charge separation within clouds. Most lightning is intracloud or intercloud; only 20%–25% of lightning strikes are cloud-to-ground.

bottom of the cloud. As they fall through the storm cloud, graupel and hail collide with uprising snow crystals. Like your feet scraping electrons from the carpet, the larger particles scavenge electrons from smaller ice crystals when they make contact. The bottom of the cloud becomes negatively charged (excess of electrons) while the top becomes positively charged.

Since opposite charges attract, negative charges at the cloud bottom want to recombine with positive charges, either within the cloud itself or on the ground. But this process can't happen immediately—air itself is a rather poor conductor of electricity. So ice particles collide again and again, charge accumulates to excessive levels, and voltage builds and builds until—BAM!—the cloud discharges in a dramatic flash of light, heat, and sound. Lightning strikes are some of the most dangerous and mesmerizing displays of nature on Earth.

As impressive as they are, however, Earth's deadly lightning strikes are little more than doorknob zaps compared to the sizzling superbolts of lightning found on Saturn.

Surprisingly, lightning is not so easy to see from outer space. Thunderstorms often originate deep within a planet's cloudy atmosphere, and sunlight can overwhelm lightning flashes on the day side of the planet. Jupiter is the only planet other than Earth where visible lightning flashes have been observed. For other planets, we must rely on radio waves to "see" lightning.

Lightning strikes emit a broad spectrum of radio waves in addition to bright visible light. If you have ever listened to an AM radio station during a thunderstorm, you probably noticed static or crackling sounds caused by lightning. These lightning-induced radio waves travel great distances—the thunderstorm can be hundreds of kilometers away and still disrupt the AM radio signal.

False-color visible images taken 75 minutes apart by the Galileo spacecraft show lightning storms on the night side of Jupiter. Moonlight from Jupiter's moon Io illuminates the night side, allowing scientists to correlate lightning flashes with particular storms. The largest storm is over 2,000 km (1,200 miles) across, about the size of Germany, France, and Spain combined. The largest Jovian lightning bolts are roughly three times more energetic than the strongest lightning flashes on Earth.

When the storm's radio waves reach the charged upper atmosphere (ionosphere), lower-frequency waves become trapped and propagate along magnetic field lines all the way to the other side of the planet. Trapped radio pulses are given cool names like "sferics," "tweeks," and "whistlers." When converted into sound by a radio receiver, the signals produce a noisy cacophony of hisses, pops, and whistles.

Whistler modes have been detected on Earth, Venus, Jupiter, and Neptune. Radio bursts of even higher frequency have been measured on Saturn and Uranus. The Solar System is alive with the sounds of lightning. No such sounds have been heard more loudly than those detected by the Cassini spacecraft at Saturn.

On its long journey to Saturn, the Cassini spacecraft received gravitational boosts as it flew around the Sun and again by Earth. During the close approach to Earth, Cassini detected radio bursts caused by terrestrial lightning at a distance of 89,000 km (55,000 miles) above the surface. At Saturn, however, radio pulses (called Saturn electrostatic discharges, or SEDs) were detected 161 million km (100 million miles) away, making the radio signal—and the lightning—about a million times more powerful.

Unfortunately, these superbolts have yet to be detected *visibly* on the night side of Saturn. Ring-shine is very bright (yes, the rings reflect light onto the night side of Saturn just like our Moon reflects sunlight onto Earth) and the water ice clouds deep in the atmosphere are usually covered with a thick haze. But Cassini scientists have

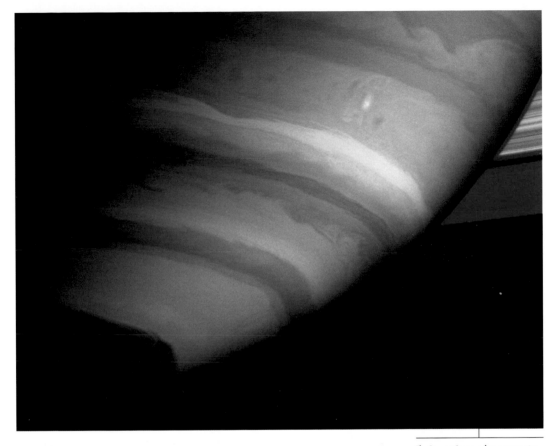

correlated SEDs with bright clouds in a region called Storm Alley at 35°S latitude. Radio bursts begin just before these clouds appear over the horizon. The repeatability of these observations leaves little uncertainty regarding the source of the SEDs.

Just imagine what Saturn's thunderstorms must be like. While lightning bolts on Earth have the thickness of a banana, Saturn's superbolts could be as wide as the Eiffel Tower. A spacecraft the size of the space shuttle could be completely engulfed by a single stroke of Saturn lightning. Without a doubt, a direct probe into a Saturn thunderstorm with electronic scientific instruments would be incredibly risky—and totally awesome (in a purely scientific sense, of course).

So the next time you get zapped by your doorknob on a cold winter's

The Dragon Storm, a large convective storm with bizarre "arms," propagates in "Storm Alley" in the southern hemisphere of Saturn. This storm was a powerful source of radio emissions, a strong indicator of lightning activity. Lightning on Saturn may be a million times more powerful than lightning on Earth.

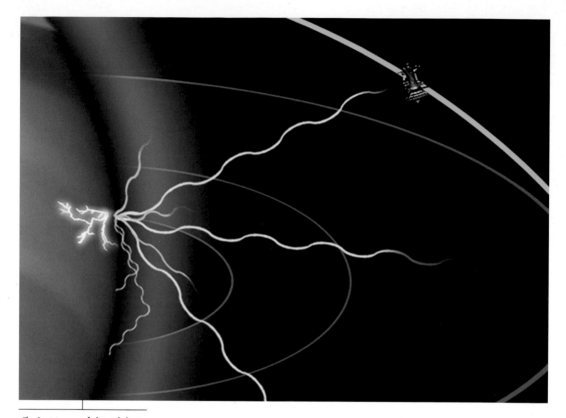

The Cassini spacecraft detects lightning at Saturn by listening for radio bursts. Lightning emits electromagnetic radiation at various wavelengths, including visible light and radio waves. Low-frequency radio waves are trapped by Saturn's ionosphere, but high-frequency radio waves can be detected by the spacecraft.

day or a radio broadcast gets disrupted by distant lightning, think about the powerful superbolts on Saturn. Why are lightning storms stronger on Saturn than on any other planet, especially more massive Jupiter? What causes energetic storms to form along Storm Alley? How can we get direct visual confirmation of these superbolts? It will take shocking new discoveries and flashes of insight to resolve these electrifying mysteries.

Part Seven

Life

Juuust Right!—Earth's Abundant Life

Okay, okay . . . there is no real competition here. As far as we know, Earth is the only place in the Solar System with living things. Yes, it is possible that microbial life exists in other places—in the subterranean ocean of Europa, in the hydrocarbon lakes of Titan, or even perhaps in suspended animation aboard a frozen comet.

But our planet *teems* with life. Everywhere you look on the surface—from the deep ocean to the highest mountains, from the hottest jungles to the frozen tundra—you see the signatures of life: self-organization, metabolism, growth, response to stimuli, adaptation, sex, poop.

Truth is, a good scientific definition of life has been extremely difficult to devise. Not all life has the same characteristics, and inanimate objects can sometimes look "alive." Fire, for example, has many characteristics of life (organization, energy conversion, growth, response to stimuli), but few people would consider it a life-form. Discoveries of new organisms, such as extremophiles (extreme-loving microbes) found in scorching geothermal springs, force us to broaden our previous ideas regarding life. Plus, a strict definition based on terrestrial life may not apply to extraterrestrials at all.

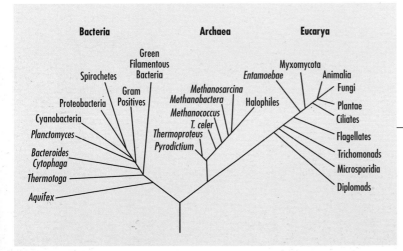

The tree of life based on genetics. All life on Earth can be categorized into three domains: Bacteria, Archaea, and Eucarya. The Bacteria and Archaea domains consist of single-celled microorganisms, while the Eucarya domain contains more complex organisms with cell nuclei. Bacteria are the most abundant organisms on Earth—a human mouth is home to more bacteria than the number of people in the world. Despite the structural similarity to Bacteria, Archaea organisms possess genes more closely related to organisms of Eucarya. All plants and animals fall within the Eucarya domain. Perhaps as many as 100 million species fill the tree of life.

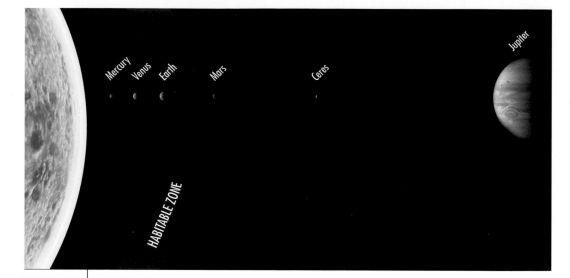

The habitable zone refers to the region in the Solar System where liquid water can exist on a planet's surface. Earth is the only planet to reside in the habitable zone. Life doesn't automatically occur within the habitable zone: the Moon is too small (and its temperature swings too large) to sustain life. Life could possibly survive outside the habitable zone, perhaps in a subsurface liquid ocean or deep within an unfrozen crust.

So even though we can't exactly define it, we've certainly got a lot of it here on Earth. Roughly 1.6 million different species have been identified, and there are many more that remain undiscovered. Every year, thousands of new species are found—including microscopic plankton, crawling fish, hairy lobsters, legless lizards, leafless orchids, and big-toothed leopards. Estimates place the total number of species on Earth at an astounding 30–100 million!

To achieve these kinds of numbers, conditions on Earth had to be just right—not too hot, not too cold—the so-called Goldilocks Principle. Since liquid water acts as a natural solvent for critical biochemical reactions, a planet with liquid water on its surface is primed to support life. There is only a narrow band in our Solar System (from 0.95 to 1.37 AU) where water is able to remain in liquid form at the surface. Earth orbits along the inside edge of this habitable zone.

Yet according to astrobiologists Peter Ward and Donald Brownlee, simply residing in the liquid water habitable zone is not enough. Myriad other factors must converge for complex life to develop. The strong gravitational influence of *Jupiter* prevents massive asteroids and comets from continually bombarding our planet. Unlike Mars or the Moon, Earth is large enough not only to keep a *substantial atmo-*

sphere that regulates temperature but also to produce a *strong planetary magnetic field* that protects against high-energy solar particles. Our large *Moon* acts like a gyroscope to stabilize the Earth's tilt and thus keeps potential climate swings in check. *Plate tectonics* allows continual recycling of carbon; without it, too much CO_2 would reside in the atmosphere and we would have a runaway greenhouse effect like Venus. All of these factors allowed life to evolve in a relatively stable environment.

Despite this lucky convergence, the diversity of life we see today has not happened quickly or without trauma. It was difficult for life to become established on Earth during the Late Heavy Bombardment period 3.8–4.1 billion years ago (hey Jupiter, what happened?). Single-cell microbes from the Bacteria and Archaea domains ruled the Earth for roughly 2 billion years, then simple multicellular organisms in the Eucarya domain reigned for another billion years or so. Complex life—including phytoplankton and small animals—exploded onto the scene only relatively recently, about 530 million years ago. Since this time, five major mass extinction events have blitzed planet Earth. These extinctions were likely induced by giant asteroid impacts, massive volcanism, or abrupt natural global warming/cooling.

We are now undergoing a sixth mass extinction, this time caused

Timeline of major events in Earth's history. Although humans have existed for only a tiny fraction of the age of the Earth, life itself may be nearly 4 billion years old.

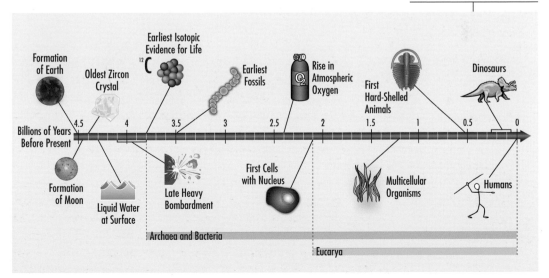

Diversity of Life on Earth: Extreme Examples

Methanococcus jannaschii, a tiny Archaea organism that feeds off methane rather than photosynthesis, was discovered at the base of a white "smoker" at the bottom of the eastern Pacific.

Thank you, bacteria! Cyanobacteria (blue-green algae from the Bacteria domain) are considered primarily responsible for putting oxygen into the atmosphere and ocean through photosynthesis 2.45 billion years ago.

Taking one for the team: the female black-and-yellow garden spider (*Argiope aurantia*) eats her male companion after mating.

Amorphophallus titanum (literally "huge shapeless male genitalia") is perhaps the tallest flower in the world, growing as high as 2.4 m (8 ft). While blooming, it exudes a smell similar to that of decaying flesh to attract flies.

The Great Barrier Reef stretches over 2,600 km (1,600 miles) along the northeast coast of Australia. It is the most massive and most diverse ecosystem on Earth.

The vicious *Felis catus,* or domestic cat, a distant cousin of the newly discovered Borneo big-toothed leopard.

Shrimplike krill, individually only a few centimeters in length, collectively account for the most biomass of any animal on Earth.

by humans. Rich, diverse ecosystems are being rapidly destroyed by deforestation. Toxic pollution creates dead zones in the ocean. Human-induced global climate change—warmer temperatures, rising sea levels, and shifting weather patterns—places undue stress on species from the tropics to the poles. Species are becoming extinct at an alarming rate—as many as three species per hour! By the year 2100, nearly half of the species on Earth may vanish.

At the end of the classic fable, Goldilocks runs away from the three bears, never to be seen again. We must work hard to ensure the same fate doesn't befall the diverse tree of life. Earth is a very special place in our Solar System—it's the "just right" home of abundant life.

It happened almost instantly, a mere blink of an eye in geological time. Plunging at supersonic speeds of 20 km/s (45,000 mph), an asteroid the size of Mount Everest targeted the shallow seas of present-day Yucatán. Upon impact, the asteroid displaced tons of water and vaporized the underlying limestone rock, gouging a crater 180 km (110 miles) wide and 35 km (21 miles) deep. A huge tsunami, perhaps taller than a 20-story building, swelled through the Gulf of Mexico and the Caribbean Sea, destroying everything in its path.

The damage was not confined to North America. Dust, water vapor, carbon, and sulfur blasted high into the atmosphere and spread around the globe. Sulfuric acid as corrosive as that found in automo-

Sixty-five million years ago, an asteroid roughly 10 km (6 miles) in diameter slammed into the shallow seas off the Yucatán peninsula, creating a large tsunami, global wildfires, and global climate change. This devastating impact is the likely cause of one of the greatest mass extinctions in the fossil record.

bile batteries rained down on Earth. The giant impact rung our planet like a gigantic bell, causing massive earthquakes and volcanic eruptions worldwide. The intense energy of the impact—a billion times greater than that of an atomic bomb—superheated the atmosphere and ignited global wildfires. The Earth was a violent, caustic fireball.

In the aftermath, the climate abruptly switched. Dust and soot in the stratosphere blocked out enough sunlight to cause perpetual night. The frigid, winterlike conditions lasted for years. Once the dust settled and sunlight returned, an intense greenhouse effect due to increased atmospheric carbon dioxide kicked in. Temperatures rose to unprecedented levels.

Only the strong—and lucky—could survive. About 70% of all life on Earth, including the dinosaurs, vanished.* Known as the Cretaceous-Tertiary event (K-T for short; the K comes from the German word for "Cretaceous," *Kreideziete*), this mass extinction occurred 65 million years ago. It signaled the end of the geologic period known as the Cretaceous and opened the Tertiary period. It is one of the most catastrophic events ever to happen to our planet.

Ironically, an earlier great mass extinction—the Permian extinction event, which wiped out 95% of all species—helped usher in the "Age of Dinosaurs." The clean biological slate, along with a warm tropical climate associated with all the Earth's landmasses being pressed into one supercontinent (Pangea), allowed dinosaurs to flourish. The giant reptiles reigned the Earth through the Triassic, Jurassic, and Cretaceous periods for over 165 million years.

But the "terrible lizards" couldn't survive the K-T impact. If the immediate destructive assault didn't get them, the ensuing darkness, which shut down photosynthesis and killed the bountiful vegetation that supported the dinosaur food chain, certainly did. Smaller mammals and birds subsequently filled the ecological niches left vacant by the dinosaur die-off. The "Age of Mammals" had begun.

How do we know that a large impact annihilated the dinosaurs? The clues lie within a thin layer of clay that marks the boundary between

---

*Although the dinosaurs are the most famous example of extinction, other important species also died, including 90% of all oceanic plankton, all ammonites (prolific marine invertebrates with nautilus-like chambers), and the large winged flying pterosaurs.

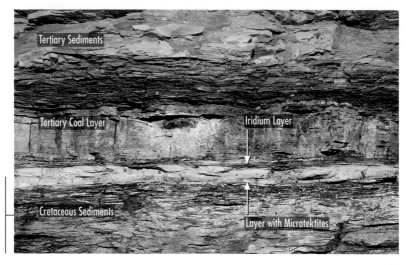

Tertiary Sediments

Tertiary Coal Layer

Iridium Layer

Cretaceous Sediments

Layer with Microtektites

The K-T boundary in Trinidad Lake State Park, Colorado, is marked by a thin layer of clay enriched in iridium, shocked quartz, and glass spherules. These features are best explained by a large asteroid impact.

the Cretaceous and Tertiary periods. Extraordinary amounts of iridium, a metal more rare than gold on Earth but relatively abundant in meteorites, are found worldwide within the K-T layer. Grains of shocked quartz, tiny glass spherules, and unusually shaped glass tektites—all telltale signs of a hot, high-pressure impact—are found within the clay layer in North America. And in many parts of the world, graphite particles likely caused by burning are detected at the K-T boundary.

To explain the presence of iridium, the father-son team of Luis and Walter Alvarez proposed in 1980 that a large asteroid impact spread iridium-laced dust globally. Furthermore, to explain the prevalence of shocked quartz and glass particles found there, their hypothesis suggested that the crater should be located in North America. The only problem: Such an impact should leave a huge, noticeable scar. No such 65-million-year-old crater had been found.

It wasn't until 1990, using proprietary oil-drilling data taken four decades earlier, that scientists confirmed a K-T-aged crater buried beneath sediments 1 km (0.6 mile) deep in the Yucatán peninsula. Named Chicxulub after the nearby town (and the Mayan word for "tail of the devil"), this feature is one of the largest and best-preserved impact basins on Earth—think of the overlying sediments as mummifying the crater and protecting it from erosion. Subsequent chemical analyses showed that melted rock in the Chicxulub crater and the K–T glass spherules from other locations around the world all came from the same source rock.

Despite this compelling evidence, the impact hypothesis is not without controversy. Borehole samples from Chicxulub place the iridium layer 300,000 years *earlier* (or 14 m/46 ft deeper in the rock) than the fossil decline. This evidence suggests that the impact could not have produced the *immediate* demise of the dinosaurs. Other scientists argue that the geologic record near the Chicxulub crater has been "reworked" by tsunamis and landslides— rocks near the crater are all jumbled up and give an inaccurate time-line of events. Samples farther away from the crater show an exact correspondence between iridium and mass extinction.

Another possibility is that a different catastrophe—volcanic activity—altered the climate and caused mass extinction. Roughly 64–67 million years ago, a supervolcano in the Deccan region of India spewed lava in volumes that dwarf modern-day Hawaiian eruptions. Intriguingly, the main pulse of this volcanism ended at around

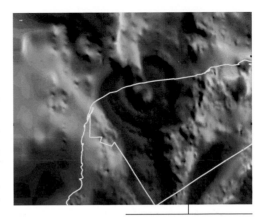

Variations in gravity reveal the Chicxulub crater. Warm colors indicate high-density rocks (high values of gravity), and cool colors lower-density rocks. The lower-gravity values within the crater are caused by sediments that filled the basin. The multiring structure is characteristic of large impacts. White lines indicate present-day boundaries of the Yucatán peninsula.

Impacts and Their Effects				
Impactor Diameter	Crater Diameter	Frequency	Example	Effect
~1,000 km	—	4.5 billion years	Moon formation	Planet's surface melted
~100 km	~2,000 km	4 billion years	Moon's Aitken Basin; none left on Earth	Ocean vaporized, life below surface may have survived
~10 km	~200 km	100 million years	Chicxulub, Mexico	Global fires, tsunami, global dust, acid rain, darkness, mass extinction
~1 km	~20 km	500,000 years	Ries Crater, Germany	Global dusty atmosphere, global ozone destruction, area size of France destroyed
~100 m	~2 km	3,000 years	Meteor Crater, Arizona	Large city destroyed, beautiful sunsets
~10 m	200 m	20 years	Boxhole Crater, Australia	Localized damage
After de Pater and Lissauer (2001)				

Landscape devastation from the
Tunguska meteoroid blast in 1908.

the same time as the K-T boundary. Perhaps the supervolcano and the asteroid tag-teamed the dinosaurs: Deccan volcanism weakened life on Earth, and the asteroid impact landed the knock-out blow.

Asteroid attacks from outer space pose an obvious risk to life on Earth. Luckily, monstrous impacts like the K-T event are extremely rare, occurring once every 100 million years or so. But smaller impacts that could destroy a city occur more regularly, roughly every 100–3,000 years. The last such event occurred in 1908 near Tunguska, Siberia. The meteoroid (perhaps a small asteroid or comet fragment) exploded in midair, creating an intense shock wave that flattened trees over an area the size of London. Hmmm, every hundred years . . . 1908 . . . Wait a minute, we're overdue!

Fortunately, several international programs are currently surveying the known asteroids to determine the likelihood of another giant impact on Earth. NASA's Near Earth Object (NEO) program alone tracks roughly 6,000 asteroids and comets in Earth's neighborhood. One recent success of the NEO program was the prediction of the impact of asteroid 2008 TC3, a spectacular fireball that streaked across the skies of northern Sudan on October 7, 2008. This was the first object to be tracked and predicted to hit the Earth prior to its actual impact.

Luckily, 2008 TC3 was relatively small (only a few meters across), and currently none of the known NEOs poses any significant threat to Earth. But this could change in the future: Jupiter could give a gravitational boost to an asteroid and send it hurtling toward Earth, or a comet from the outer Solar System could become Earth crossing.

Rest assured, NASA, ESA, and others will continue to monitor the situation closely. We would *all* like for the "Age of Mammals" to last a good while longer.

Life from Above—Alien Origins

One of the downsides of having a big brain is that we continually wrestle with the origin of our existence. Bacteria, on the other hand, don't seem to struggle with these deep questions in the same way. Fortunately, Darwin's theory of evolution explains quite well how species *change* over time—humans are the result of a long line of genetic adaptation. But Darwin's theory does not address the ultimate question: how did life begin on Earth?

The surprising scientific answer—life on Earth may be the result of heavenly intervention: alien organic material brought to Earth by asteroids and comets.

In the beginning, it wasn't easy for life to become established on Earth. Persistently bombarded with high-energy planetesimals, the nascent Earth was a seething, molten cauldron. It was hot! Sustaining life during these early stages would have been virtually impossible. Large impacts, such as the one that formed the Moon, repeatedly sterilized the entire planet.

Comets with relatively high concentrations of organic material could have been the bringers of life to Earth.

Life could maintain a foothold only after the planet's surface cooled sufficiently and the frequency of the largest impacts diminished. The earliest evidence of life comes from carbon isotopes in Greenland rock layers estimated to be 3.83 billion years old. Living organisms preferentially use the lighter ^{12}C isotope (carbon atom with 6 protons and 6 neutrons) over heavier ^{13}C (6 protons, 7 neutrons). In the Greenland rocks, the $^{12}C/^{13}C$ ratio is higher than normal, indicating the possible presence of life. And the timing is right—the age of these carbonaceous layers occurs near the end of the Late Heavy Bombardment period.

Yet for life to begin, the essential ingredients must have been already present. The everyday processes of life involve stringing

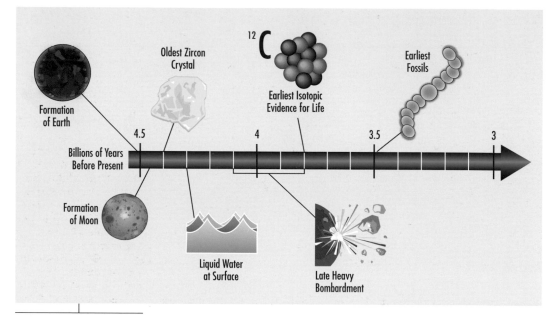

Timeline of major events in Earth's early history.

together amino acids—complex molecules of carbon, oxygen, nitrogen, and hydrogen—to create proteins. While DNA may be the blueprint of life,* proteins are the workers in a living cell. Cell structure, movement, communication, repair, and replication all depend on proteins of different sizes and shapes. So before life could begin, amino acids—the building blocks for proteins—must have been floating around.

Where did these amino acids come from? Perhaps they were present in the original cosmic soup of the young Earth. Amino acids could have formed within the hot crust and been injected into the cooler ocean through chemically active hydrothermal vents. Single cellular life originally may have been born in the dark abyss of the deep ocean.

Or perhaps the building blocks of life came from the stormy skies of the early Earth. In their famous 1952 experiment, Stanley Miller and Harold Urey showed that electrical discharges in gases of the ancient

---

*At least for life as we know it today. It is possible that early life may have been based on RNA rather than DNA, but that DNA cells took over the world because the double-helix structure of DNA is more stable than the single-helix structure of RNA.

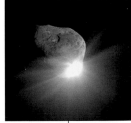

The organic recipe for comets: when the Deep Impact projectile collided with Comet 9P/Tempel (above), both the main Deep Impact spacecraft and the Spitzer Space Telescope measured the chemical signature of the resulting debris cloud. The main ingredients (left) are: on plates, from left to right—ice and dry ice; in measuring cups, from left to right—olivine, smectite clay, polycyclic aromatic hydrocarbons (organic material!), spinel, and metallic iron; on table in the front, from left to right—the silicate enstatite, the carbonate dolomite, and the iron sulfide marcasite.

atmosphere could create amino acids. Life (or at least the building blocks of life) could have been zapped into existence by lightning!

But another hypothesis for life's origins maintains that amino acids and other organic material came from outer space.* Although asteroids and comets are often thought of as harbingers of doom (an asteroid impact did wipe out the dinosaurs, after all), they may also be the ultimate bringers of life to Earth. These small Solar System bodies contain copious amounts of organics, and laboratory tests on amino acids show that they can remain intact during fiery collisions with Earth. The Murchison meteorite alone contains over 100 different amino acids and nucleobases (the foundations of RNA and DNA). Carbon analysis of this meteorite shows that much of the organic material cannot be due to Earth contamination—these organics are extraterrestrial.

Two recent NASA missions offer further evidence for organics from space. When the impactor from the Deep Impact spacecraft slammed into Comet 9P/Tempel at over 37,000 km/hr (23,000 mph), it created a massive cloud of ejecta. The debris cloud contained

---

*Interstellar space is likely littered with amino acids that materialize in dense gas clouds and star-forming regions.

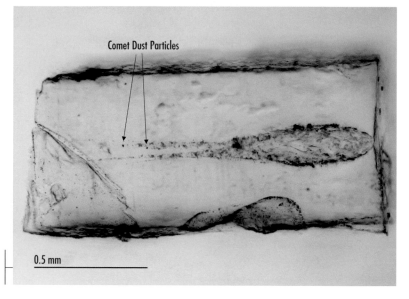

Comet Dust Particles

0.5 mm

Dust particles from Comet 81P/Wild captured in aerogel in the first ever sample return of cometary material.

the usual suspects of volatiles (water and carbon dioxide) and rock (carbonates, clays, silicates, and iron), but it also contained significant amounts of organic hydrocarbons. Full-fledged amino acids, however, were not detected during the Deep Impact crash.

The Stardust mission returned more compelling results. Using state-of-the-art aerogel, this spacecraft trapped dust particles from the coma of Comet 81P/Wild as it flew by the comet in 2004 and returned these samples to Earth for analysis. The samples contained over 10% organic material, including some amino acids! In 2009, NASA scientists reported that Stardust samples of the amino acid glycine contained more ^{13}C than is found in glycine formed on Earth, the first *direct* confirmation of extraterrestrial amino acids from a comet.

To take this idea a step further, actual life itself—not simply the building blocks of life—may have come from outer space. The panspermia hypothesis (Latin for "all life") asserts that the "seeds" of life are spread throughout the universe and can survive interplanetary—even interstellar—journeys. Most scientists today consider widespread panspermia unlikely, but new research on astrobiology shows that, remarkably, life can indeed survive in quite extreme environments.

Consider, for example, the tiny organism *Chryseobacterium green-landensis* (Greenland bacterium). This microbial life lay dormant for 120,000 years in glacial ice buried 3.2 km (2 miles) below the surface. It quickly awoke from hibernation when the ice thawed. The same could be true for microbial spores hitching a ride in asteroids and comets.

Delivery of organic material from outer space is happening today. Every year, roughly 40,000 tons of interplanetary dust—each grain smaller in diameter than a human hair—rains down upon our planet. Within this interplanetary dust, NASA researchers recently detected organic molecules that originate from the early Solar System. We are continually being bombarded by ancient aliens (well . . . at least some very old organic material).

If you think of today's cosmic rain of organics as a drizzle, the young Earth likely experienced heavy organic floods from above. The inner Solar System was filled with small chunks of primordial debris. Imagine a deluge of organic matter on our planet, with amino acids from one meteorite mingling with amino acids from another. Perhaps by accident, these amino acids combined into proteins, and then these proteins twisted and folded in special ways. During this turbulent dance, a protein possibly split into two but then quickly repaired itself. The magical process of life modestly—and significantly—began on Earth.

Little Green . . . Microbes?—Possible Life on Mars

Science progresses in fits and starts of discovery and rethinking. Unlike many ways of knowing, science requires empirical evidence—not just belief—to explain the natural world. Carl Sagan, one of the premier planetary scientists of the 20th century, claimed that a scientist's job is to make as many hypotheses as possible and then try to disprove them with solid evidence. If a hypothesis repeatedly survives, it may be elevated to the level of theory. Scientific theories stand the test of time by explaining both old observations and new discoveries. Yet often, technological advances reveal new evidence that causes long-standing theories to be thrown out. Nowhere is this process more apparent than in the search for life on Mars.

For centuries, the planet Mars was considered to be nothing like Earth, a distant bloodred wanderer in the night sky. However, telescopic observations in the 18th century showed bright features, dark regions, and ice caps reminiscent of those on Earth. These features fueled speculation of possible life on Mars.

In 1877, Giovanni Schiaparelli observed long straight features over

Comparison of 1894 sketch of Mars by Eugene Antoniadi with 2003 Hubble Space Telescope image. Although large-scale features are accurately observed in Antoniadi's sketch, the extensive canal system constructed by Martians (as suggested by Percival Lowell) does not exist. Antoniadi later recognized the canal network as an optical illusion after making observations with a more powerful telescope. South is toward the top in this image.

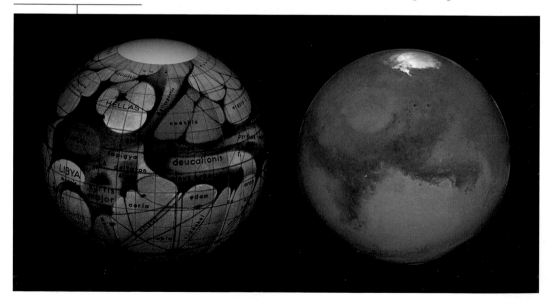

the entire planet. He called these features *canali* (Italian for channels), which were unfortunately translated into English as "canals."
Schiaparelli's observations motivated businessman Percival Lowell
to establish an observatory in Arizona in 1894 primarily to investigate
life on Mars. Lowell hypothesized that the intricate canals were built
by an ancient intelligent martian civilization. Lowell's work inspired
H. G. Wells's 1898 novel *War of the Worlds,* in which Martians invade
Earth to escape their dying planet.

Can science save us from the Martians
(or at least from bad movies)?

Technological advances again altered the prevailing scientific
theory in the early 20th century. Like Lowell, Greek astronomer
Eugene Antoniadi had observed the canal structure on Mars for
many years. During the 1909 Mars opposition (closest approach
to Earth), Antoniadi used the new large 83-cm telescope at the
Paris Observatory to make detailed observations of the planet.
Surprisingly, Antoniadi saw no canals on Mars! He concluded that
the canals were an optical illusion produced at the viewing limits of
smaller telescopes.

Early spacecraft missions to Mars dealt another blow to speculation of extraterrestrial life on Mars. In 1965, the Mariner 4 spacecraft
returned images of the cratered martian surface devoid of canals,
flowing water, and vegetation. Two Viking spacecraft conducted four
different biological experiments on soil samples from the martian
surface in 1976. The experiments revealed a sterile environment
with minimal amounts of organic material. Even the one experiment with a positive result, the Labeled Release (LR) experiment in

Trench marks indicate the location
of soil samples taken for Viking
Lander 1 biological experiments. The
digging tool (lower center) would
scoop up samples and place them on
the lander for analysis. The Viking
experiments found no life on Mars.

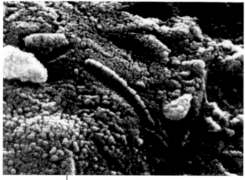

Martian meteorite ALH84001 (top) and a close-up scanning electron microscope image within the meteorite (bottom) of a wormlike structure interpreted by some scientists to be a microfossil of ancient nanobacteria. Nonbiological activity can also explain the presence of this structure.

which organisms (if present) would digest radioactively tagged nutrients and subsequently burp radioactive gas, can be more readily explained by nonbiological breakdown of the nutrients. The search for life on Mars essentially stalled.

Then in 1996, NASA scientists made a startling announcement—analysis of a meteorite from Mars suggested possible ancient microbial life on the Red Planet. Scientists discovered the meteorite, named ALH84001, during a 1984 expedition in Allan Hills, Antarctica. It likely arrived in Antarctica after being blasted from the martian surface by asteroidal or cometary impact.

There were three lines of evidence for possible life: 1) the presence of polycyclic aromatic hydrocarbons (PAHs), organic molecules often produced by decaying microorganisms, 2) unusual carbonate globules with magnetite similar to those produced by Earth bacteria, and 3) tiny wormlike blobs with a striking resemblance to microfossils on Earth, albeit 100 times smaller. All of these features were found within a few microns of one another (a pinhead is about 2,000 microns in diameter).

This extraordinary claim sparked intense research on ALH84001. Plausible nonbiological explanations exist for each line of evidence, and PAHs within the meteorite are more likely due to contamination from Antarctica ice. Today, most planetary scientists remain unconvinced that ALH84001 contains proof for past martian life. Nevertheless, the ALH84001 discovery spawned the exciting new field of exobiology—the study of life on other planets.

We don't know if Mars ever supported life or if it harbors life today. Contradictory evidence clouds the issue, and our understanding of life is colored by our experience on Earth. Until recently, scientists believed that life required temperate climates for survival. However, the discovery of extremophiles, microbial organisms that live in extreme conditions such as hydrothermal vents on the ocean

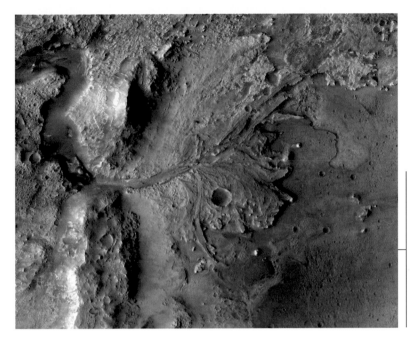

New observations from the state-of-the-art Compact Reconnaissance Imaging Spectrometer for Mars (CRISM) offer strong evidence for past water on Mars. Here, an ancient river channel flows into the dry lake bed of Jezero Crater. The delta region contains abundant clay-bearing minerals (green), which tend to trap organic material. Jezero Crater is a prime location to search for ancient martian life.

floor, rekindled the possibility of finding life in the harsh martian environment.

There are some tantalizing possibilities. Recent investigations show that the primitive technology of the "definitive" Viking biological experiments cannot detect life even in biologically active soil from Antarctica and Chile's Atacama Desert. In 2008, the Mars Phoenix Lander "tasted" water (the elixir of life as we know it) as it melted frozen soil from the martian arctic in one of its ovens. And variations in atmospheric methane detected by the Mars Express spacecraft could be caused either by volcanic activity or by microbes. If life does exist on Mars, it will most likely be microbial rather than advanced.

What a remarkable and humbling day it will be if life, however small, is found on the Red Planet. It will mark a milestone in humanity's scientific quest for knowledge. Yet at this point, this is simply an unproven hypothesis. Even more discoveries triggered by new technologies—such as those engineered for NASA's Curiosity (2011) and ESA's ExoMars (2018) missions—will be required to answer definitively: Is there life on Mars?

Life in the Dark—Earth & Europa?

Life without light . . . The Endeavour Hot Vents, at a depth of 2,200 m (7,300 ft) near the Juan de Fuca Ridge, gush out mineral-rich hot water. Without the benefit of photosynthesis, giant tube worms obtain their life-supporting energy via a symbiotic relationship with sulfur-eating bacteria.

The scene below them was totally unexpected. In 1977, scientists aboard the deep-sea submersible *Alvin* were searching for hydro-thermal vents along the Galapagos Rift roughly 2.5 km (1.6 miles) beneath the sea. They wanted to confirm geothermal activity at seafloor spreading ridges as predicted by plate tectonics theory. In the murky deep, scientists anticipated a barren wasteland of fresh lava flows and bubbly vents.

What they found instead forever changed the way we view life on Earth. It also revealed exciting possibilities of how life may flourish in extreme environments beyond our home planet.

Led by explorer Robert Ballard (who would later use *Alvin* to search the sunken ocean liner *Titanic*), the expedition did find hydrothermal vents—"black smokers" that spew superheated water (~375°C/710°F) into the frigid dark ocean. Liquid water is able to exist at brick-oven temperatures because of the immense pressures at the sea bottom—the boiling point is increased to about 400°C (750°F) at the Galapagos Rift. Hot water from the crust is enriched in sulfides, producing the black "smoke" emanating from the vents.

Surrounding the hot chimneys was an exotic ecosystem unlike anything seen before. . . . Dense colonies of snow white clams firmly attached to the toxic vents. Funky dandelion-like animals scavenging the ocean bottom. Giant tube worms with distinctive red tips swaying in the shimmering water like wildflowers in a meadow. A strange purple octopus silently preying on brown mussels.

This bizarre world exists completely without sunlight. Plants at the surface store energy through photosynthesis—sunlight helps convert carbon dioxide and water into carbohydrates. But solar radiation can penetrate only to depths of about 300 m (1,000 feet). Before 1977, scientists believed that any life surviving at the dark ocean bottom must feed on dead plants and animals, originally supported by photosynthesis, falling from the surface.

Hydrothermal vent and coral reef ecosystems intermingle at an underwater volcano in the western Pacific. At only 190 m (630 ft) below the surface, diffuse sunlight provides enough energy to support photosynthetic red algae and coral, while hydrogen sulfide from hydrothermal vents supplies chemosynthetic energy for white bacteria mats.

Extremophiles have been found in harsh environments around the globe: acidic geothermal pools, dark hydrothermal vents on the ocean floor, bone-dry deserts, and frozen lakes in Antarctica.

Life at hydrothermal vents is different, *extremely* different. Rather than using sunlight, it survives on energy from the Earth's interior.

Take the giant tube worms, for instance. Bacteria in the guts of tube worms combine hydrogen sulfide,* oxygen, and carbon dioxide into complex organic molecules (and excrete white sulfur filaments as waste). Tube worms then digest the organic molecules as food. This process of chemosynthesis—production of carbohydrates through chemical reactions without sunlight—provides the foundation of the entire food chain at hydrothermal vents.

Extremophiles, or extreme-loving organisms, also thrive in harsh places other than deep-sea vents. Some extremophiles depend on photosynthesis, and some on chemosynthesis, but all live in environments that would have been unthinkable a few decades ago. Yellowstone National Park is alive with single-celled thermophiles (heat lovers) that inhabit sulfur-riddled geothermal pools. The Atacama Desert in Chile—the driest place on Earth, so dry that NASA uses it to test Mars rovers—supports microbes that go in and out of hibernation as minuscule amounts of water become available. And

---

*Hydrogen sulfide, H_2S, is very similar in chemical structure to water's H_2O.

50 μm

They can bear it: tardigrades can live in temperatures from –200°C to 150°C (–328°F to 300°F) and pressures from a vacuum to over 1,000 atmospheres. Tardigrades have even survived direct exposure in outer space.

colonies of bacteria have been discovered in refrozen lake ice just above Lake Vostok, a subglacial lake buried roughly 4 km (2.5 miles) under the Antarctic ice shelf.

Then there are the cute tardigrades, tiny invertebrates commonly known as "water bears." They are probably living in a stream near you. Tardigrades are virtually indestructible . . . you can freeze them, boil them, zap them with intense radiation—and they still survive! In 2007, tardigrades were launched on a European Space Agency satellite and exposed to the hazards of space travel: cold vacuum, cosmic radiation, and solar radiation. Low pressures and cosmic rays left tardigrades unharmed, and some tardigrades even recovered from deadly ultraviolet radiation that would have killed most life-forms instantly.

If extremophiles can flourish in severe environments on Earth (and space), then they certainly may live in other extreme places in our Solar System. One intriguing possibility is Jupiter's moon Europa. This large icy satellite has a deep water ocean covered by a global layer of ice. Although other icy bodies may have oceans, only Europa's ocean is in direct contact with a rocky mantle (oceans on the other planetary bodies are likely sandwiched between ice layers).

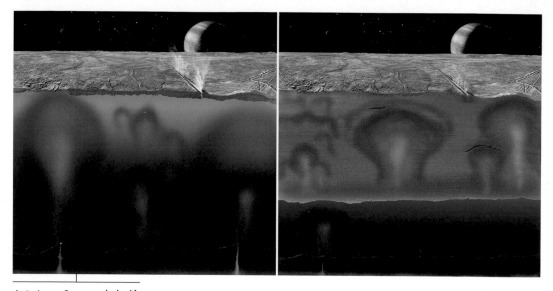

Jupiter's moon Europa may harbor life in its deep ocean. The exact structure of Europa's ocean is a current area of research. It may have a thin ice "crust" with life in the deep ocean (like hydrothermal vents on Earth), or it may have a thicker ice crust with relatively warm deep ice that may support extreme microbial life (like Lake Vostok in Antarctica).

The mantle could provide the necessary nutrients for chemosynthesis. NASA scientists are currently planning a mission to Europa to look for signs of life.

Yet when searching for life beyond Earth, we really are "in the dark." We don't know what alien life might be like. Our experience on Earth leads us to expect carbon-based life-forms—we don't know of any other kind. Perhaps alien life will be silicon-based. Possibly it will depend on genetics very different from DNA. Maybe weird alien organisms don't need water at all. One thing is certain . . . if and when (we think when) life is discovered on other planetary bodies, it will be truly enlightening.

Part Eight

The Wack Pack

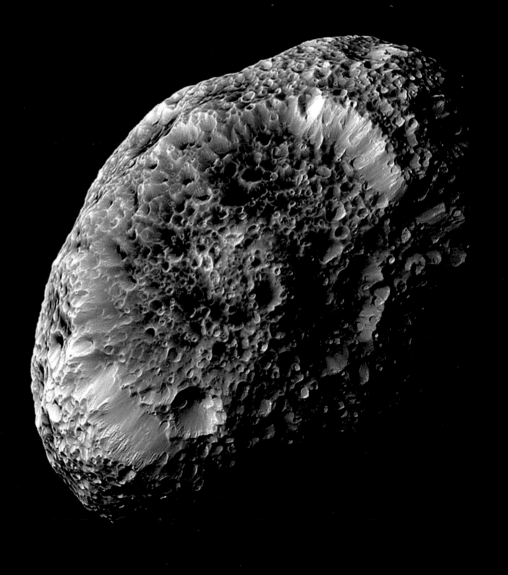

Stinkiest Place—The Rotten Egg of Io

What makes a place stinky? It is quite subjective, really. Many scientists believe that smells warn us of potentially harmful chemicals—bad smells indicate things that are bad for us (although this theory does not explain why humans eat liver and onions). However, many things stinky to humans are beneficial to other living things. Rats inhabit putrid trash bins, some bacteria eat methane, and flies like poop. So, we have taken a distinctly human perspective to determine the stinkiest place in the Solar System. It's really no contest—Jupiter's moon Io produces the biggest stink of all.

This is not to say that the Solar System lacks stinky places. Earth has its fair share—boggy marshes, geothermal areas, landfills, and

Io—the "funky" moon of Jupiter. Numerous sulfur-belching volcanoes dot Io's mottled surface, making it not only the most volcanically active body in the Solar System but also the stinkiest.

pig sties, to name a few. New Zealand's 55 million sheep and cattle produce about 90% of the country's methane emissions (Kiwi farmers have fervently opposed a "flatulence" tax). Indonesia is home to the exotic and popular durian, a fruit with such an offensive odor that it is banned in Singapore hotels and subways. On Venus, corrosive clouds of sulfuric acid completely envelop the planet, producing an acrid funk of global proportions. Hydrocarbons on Saturn's moon Titan fall as toxic precipitation, pool in lakes and streams on the surface, and seep into the underlying goopy soil. Titan smells like one big oil refinery.

But Jupiter's moon Io smells like a jumbo rotten egg. Hydrogen sulfide produces the characteristic rotten egg stench, and the stinky compound has been observed both on Io's surface and in the upper atmosphere. In fact, the sheer abundance of sulfur compounds creates the moon's distinctive red and yellow coloration.

Sulfur erupts from Io's volcanoes in gaseous molecular form S_2. These molecules combine to produce a red-colored frost on the surface. Over time, sulfur molecules form a stable ring configuration with a typical sulfur yellow color.

During explosive volcanic eruptions—spectacular events that are quite common on Io—sulfur gas launches high into the atmosphere. Although the mechanism is not completely understood, photochemistry decomposes some of this sulfur gas and forms sulfur dioxide, hydrogen sulfide, and other sulfuric compounds. Sulfur molecules freeze on the surface to form red patches near volcano vents. Eventually, the red frost turns yellow as sulfur molecules react to form stable sulfur rings.

NASA's New Horizons spacecraft captured Io in its full volcanic glory in 2007: a giant plume from Tvashtar Volcano (top), a smaller eruption on the limb from Prometheus Volcano (left center), and a bright volcanic plume from Masubi Volcano on the night side (bottom).

Like most persistently stinky places, Io continuously gets a fresh (or not so fresh) supply of malodorous compounds. It helps that Io is the most volcanically active place in the Solar System.

The sulfur-covered moon earns this di-stink-tion by traveling dangerously close to its parent planet, Jupiter, in a slightly elliptical orbit. The (rotten) egg-shaped orbit is maintained by resonances with two of Jupiter's largest moons, Europa and Ganymede: Io completes four orbits in the same time that Europa completes two and Ganymede one. Every seven days, the perfect alignment of the large satellites gives Io an extra gravitational boost that prevents the stinky moon from developing a nice circular orbit.

As Io repeatedly moves closer to and farther away from the giant planet, Jupiter's powerful gravity flexes and deforms the nearby satellite. If you quickly bend a short length of stiff metal wire (such as a wire clothes hanger) back and forth, you'll get some idea of just how Jupiter's tidal flexing generates smoldering heat deep within Io's interior.

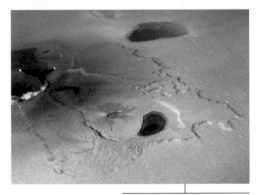

A close-up view of Tvashtar caldera by the Galileo spacecraft in 2000. New hot lava appears white/orange on the far left. This image spans about 250 km (155 miles) across.

As a result, extreme volcanic activity permeates the tortured moon. Gaseous burps emanate from over 400 active volcanic pores on its ever-changing surface. In 2000, the Galileo spacecraft detected the birth of a new volcano on Io. Then in 2007, NASA's New Horizons mission observed the largest eruption to

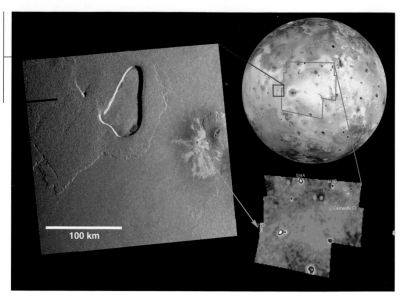

In 2000, the Galileo spacecraft detected a new shield volcano and bright surface flows with both visible (left) and near-infrared (lower right) imagery. Io glows in the infrared with volcanic "hotspots" pervading the large moon's surface.

date—sulfur plumes extending as high as 300 km (180 miles) above the surface. By comparison, the biggest volcanic eruptions on Earth reach only about 20 km (12 miles) high.

There is no indication that this overwhelmingly pungent process will end anytime soon. Jupiter keeps pulling and Io's volcanoes continue to erupt. The Sun incessantly zaps the sulfurous atmosphere with radiation. And the stinkiest place in the Solar System ends up reeking like a cracked rotten egg again and again.

Best Fuel Depot—Titan

Fed up with soaring gasoline prices? Looking for an abundant source of fuel? Look no further than Titan, Saturn's largest moon.

Titan is perhaps the Solar System's most unique moon. Bigger than the planet Mercury, Titan is the second largest satellite in the Solar System (barely edged out by Jupiter's moon Ganymede). However, Titan is the only satellite with a substantial atmosphere. Because Titan is almost 10 times farther from the Sun than Earth and nearly twice as far as Ganymede, the average surface temperature remains a chilling –179°C (–290°F). Heavy gases like nitrogen and methane lack the thermal energy to escape Titan's strong gravitational pull. This allows a dense atmosphere to enshroud the large moon. The atmospheric surface pressure on Titan is 60% higher than on Earth,

Smoggy Titan passes behind the rings of Saturn in this image from the Cassini spacecraft. The small moon Epimetheus is in the foreground.

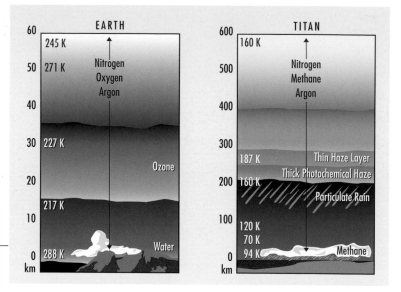

EARTH

km		
60	245 K	
50	271 K	Nitrogen Oxygen Argon
40		
30	227 K	
20		Ozone
10	217 K	
0	288 K	Water

TITAN

km		
600	160 K	
500		Nitrogen Methane Argon
400		
300	187 K	Thin Haze Layer
200	160 K	Thick Photochemical Haze
		Particulate Rain
100	120 K / 70 K	
0	94 K	Methane

The nitrogen-rich atmospheres of Earth and Titan both exhibit major hydrological cycles, water on Earth and methane on Titan. Note that Titan's atmosphere extends 10 times higher into space.

The Huygens probe detected river channels flowing into a "dry coastline."

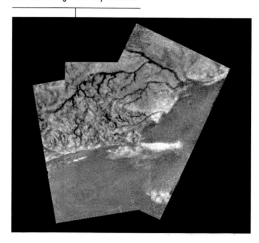

similar to the pressure you would feel diving to a depth of 6 m (20 ft) in the ocean. Since the atmospheres of both Earth and Titan are mostly nitrogen, Titan may mirror conditions on primitive Earth before life emerged.

Yet observations of Titan's surface from Earth have proved extremely difficult. Photochemical smog fills Titan's atmosphere, obscuring the surface with a thick haze. One of the main goals of the current Cassini-Huygens mission at Saturn is to unlock the secrets of Titan. The Cassini spacecraft employs a haze-penetrating radar to view the moon's surface, and in January 2005, the Huygens probe plunged through Titan's thick atmosphere to land on the surface. New results from this mission have been nothing less than astounding.

Methane, the main component of natural gas, permeates the large moon. Like water on Earth, methane on Titan exists in three phases—solid, liquid, and gas. Roughly 5% of the dense atmosphere consists of methane and other hydro-

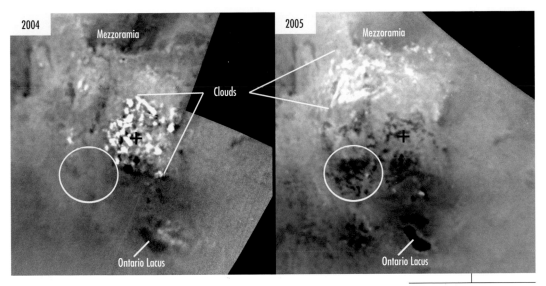

The methanological cycle at work! These images of Titan's surface near the south pole show dark areas swelling in size (see white circle). These dark features are most likely hydrocarbon lakes filled by seasonal rains of liquid methane. Clouds were frequently observed near the pole during this period. Some features in these images, such as Ontario Lacus, show changes in brightness due to differences in illumination.

carbons. Liquid methane clouds may produce raindrops larger than those on Earth, possibly generating flash-flooding events that rapidly carve the water-ice and solid-hydrocarbon surface. As the Huygens probe parachuted through Titan's atmosphere, it observed deep river channels and eroded drainage basins. The mushy landing site oozed organic compounds as the warm probe evaporated liquid hydrocarbons from the surface. In addition, Cassini spacecraft images show hundreds of methane and ethane lakes dotting the landscape.

Because of its intense methane cycle, Titan has an enormous reservoir of hydrocarbons. The largest lakes, comparable in size to Lake Superior, contain more liquid hydrocarbons than Earth's entire oil reserve. Likewise, dark organic sand dunes near the equator exceed Earth's coal reserves by 100 times or more. You might even say that the giant moon is one, er ... *Titan*ic fuel tank.

Cassini recently made another intriguing hydrocarbon discovery. In 2007, negatively charged hydrocarbon particles were detected in the upper atmosphere. These heavy negative ions may possibly allow other organic molecules to attach and form the basic building blocks for life.

If life were to exist on Titan, it would need to be dramatically different from life on Earth today to survive the harsh temperatures

Synthetic aperture radar on the Cassini spacecraft penetrates the hazy atmosphere to detect lakes of liquid ethane near the north pole. The width of the image spans about 140 km (85 miles).

and incessant methane rain. Indeed, understanding hydrocarbon processes on Titan may give valuable insight on how life first developed on our planet 3.8 billion years ago—when the early Earth's nitrogen-rich atmosphere lacked oxygen and had a persistent organic haze.

Assuming we continue to use hydrocarbons for rocket fuel, Titan will make a great pit stop during our future exploration of the outer Solar System. But Titan may serve as a potent fuel depot of another kind—a slough of hydrocarbons that could ignite into life!

Problematic Planethood—Pluto

Poor Pluto has never really fit in with all the bigger kids in the neighborhood. Something has always been just a bit different.

Ever the misfit, Pluto is smaller and icier than the rocky inner planets (Mercury, Venus, Earth, Mars), yet it travels out past the orbits of the outer gas giants (Jupiter, Saturn, Uranus, Neptune). Pluto's highly elliptical (egg-shaped) orbit is also quite a bit tilted compared to the rest of the planetary pack. This weird orbit sometimes even permits Pluto to wander closer to the Sun than does Neptune.

Although accepted as our Solar System's ninth planet after its discovery by Clyde Tombaugh in 1930, Pluto is so small and distant that it has been difficult to get a handle on its true nature. An accurate estimate of Pluto's mass couldn't be made until the discovery of

The Pluto system as captured by the Hubble Space Telescope on February 15, 2006. Currently, Pluto is defined as a dwarf planet while Charon, Hydra, and Nix are satellites of Pluto. Since Pluto is a dwarf planet as well as a Trans-Neptunian Object, the IAU definitions make it a member of the new class of Solar System bodies called plutoids.

RESOLUTION B5
Definition of a Planet in the Solar System

Contemporary observations are changing our understanding of planetary systems, and it is important that our nomenclature for objects reflect our current understanding. This applies, in particular, to the designation "planets". The word "planet" originally described "wanderers" that were known only as moving lights in the sky. Recent discoveries lead us to create a new definition, which we can make using currently available scientific information.

The IAU therefore resolves that planets and other bodies, except satellites, in our Solar System be defined into three distinct categories in the following way:

(1) A planet[1] is a celestial body that
    (a) is in orbit around the Sun,
    (b) has sufficient mass for its self-gravity to overcome rigid body forces so that it   assumes a hydrostatic equilibrium (nearly round) shape, and
    (c) has cleared the neighbourhood around its orbit.

(2) A "dwarf planet" is a celestial body that

    (a) is in orbit around the Sun,
    (b) has sufficient mass for its self-gravity to overcome rigid body forces so that it assumes a hydrostatic equilibrium (nearly round) shape[2],
    (c) has not cleared the neighbourhood around its orbit, and
    (d) is not a satellite.

(3) All other objects[3],except satellites, orbiting the Sun shall be referred to collectively as "Small Solar System Bodies".

---

[1] The eight planets are: Mercury, Venus, Earth, Mars, Jupiter, Saturn, Uranus, and Neptune.
[2] An IAU process will be established to assign borderline objects to the dwarf planet or to another category.
[3] These currently include most of the Solar System asteroids, most Trans-Neptunian Objects (TNOs),comets, and other small bodies.

its moon Charon in 1978, nearly half a century after the discovery of Pluto itself. The moons Hydra and Nix were only recently discovered in 2005 using the Hubble Space Telescope.

To make Pluto feel even worse about itself, new bodies are continually being found—in the region of the Solar System outside of Neptune's orbit known as the Kuiper Belt—that are of similar size to Pluto, or even larger. Indeed, Pluto seems to have more in common with these Kuiper Belt objects (KBOs) than with the other eight planets.

And so, after much debate, the International Astronomical Union (IAU) has made it official: Pluto is not a planet! In a controversial decision, the IAU adopted a new definition of a planet in our Solar System. In this definition, a planet must orbit our Sun, be big enough so that it assumes a mostly spherical shape, and be the dominant body in its orbit. Pluto comes up just short of being a planet with this third requirement. The diminutive body accounts for less than a 10th of the mass of all the bodies in its orbit. Earth, by comparison, has around 1.5 million times the mass of the rest of the material in its orbit.

Although Pluto's highly eccentric orbit sometimes takes it inside Neptune's orbit, it lies, on average, farther out than Neptune. This makes Pluto a member of a class of objects known as Trans-Neptunian Objects or TNOs. Any Solar System body whose average orbit lies outside that of Neptune is a TNO. The trans-Neptunian region is

The plane of Pluto's orbit is inclined (tilted) about 17° compared to the orbit of the Earth. It is also much more eccentric (noncircular) than the orbits of the other major planets. The orbit of Neptune, at about 30 AU from the Sun, marks the inner edge of the Kuiper Belt, which extends out to around 55 AU.

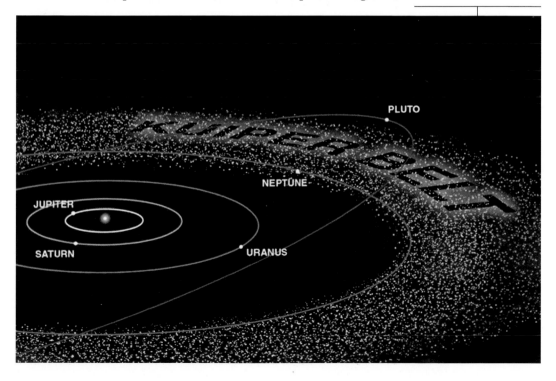

Largest known Trans-Neptunian Objects (TNOs)

Dysnomia

Charon

Eris

Pluto

Makemake

Haumea

Sedna

Orcus

Quaoar

Varuna

Size comparison of some of the largest TNOs. Quaoar (discovered in 2002) is about half the size of Pluto, Sedna (discovered in 2004) is closer in size to Pluto, and Eris (discovered in 2005) is larger than Pluto.

further subdivided into the Kuiper Belt, the Scattered Disk, and the Oort Cloud.

It was the 1992 discovery of the first TNO other than Pluto or Charon—designated (15760) 1992 QB_1—that validated the concept of the Kuiper Belt and marked the beginning of Pluto's slide from planethood. Since 1992, more than 1,000 TNOs have been found. But the 2003 discovery of Eris, a Scattered Disk object even larger than Pluto, prompted the IAU finally to reconsider the definition of a planet. Pluto didn't make the cut.

Pluto is now one of a growing number of dwarf planets. Unlike a planet, a dwarf planet does not have to sweep its orbit clear of large objects. By this definition, our Solar System currently has five known dwarf planets: Eris, Pluto, Makemake, Haumea, and Ceres (the largest object in the asteroid belt between the orbits of Mars and Jupiter).

Then in June 2008, the IAU established yet another classification: dwarf planets that are also TNOs are now designated *plutoids*. So let's see if we can get this straight. Pluto is *not* a planet. It is, however, a dwarf planet, a Kuiper Belt object, a Trans-Neptunian Object, and a plutoid. . . . Whew!

Confused yet? Well, don't try too hard to keep it straight. The controversy over the definition of a planet, both scientific and emotional, ensures that the issue of Pluto's planetary status will be revisited. In any event, Pluto should no longer feel like a misfit. After all, it's now the leader of a whole new pack of Solar System objects: plutoids!

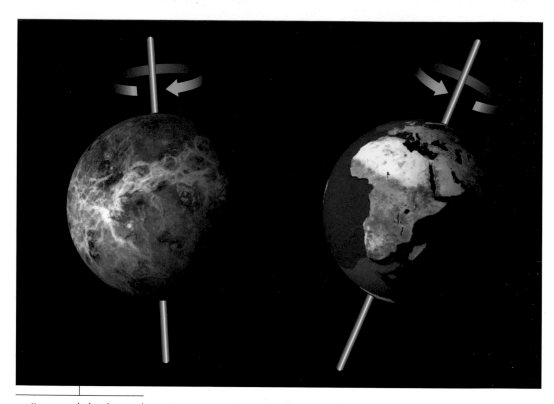

Venus rotates backward compared to most other planets (including Earth) and does so very slowly. In fact, one Venus day is *longer* than one Venus year!

When we say that the planet Venus and Neptune's moon Triton are the "most retro" in the Solar System, we don't mean they go around wearing tie-dyed shirts and listening to eight-track tapes. We're referring to fundamental characteristics of their dynamics. Both exhibit *retrograde* motion—motion opposite to the usual sense—in one fashion or another.

For Venus it is retrograde rotation. That's right, Venus rotates backward! Whereas most other planets, including Earth, rotate about their axes in the *prograde* or *direct* sense (counterclockwise when looking down from the north), cloud-cloaked Venus goes her own way, rotating clockwise—albeit very slowly.

"What do you mean by 'down from the *north*'?" you may ask. Good question! The IAU, the official body in charge of defining such things, defines a planet's north pole as the "rotational pole of a planet or satellite which lies on the north side of the invariable plane." Yes ... we hear your next question. Just what is this invariable plane and how do you find *its* north side?

Think of it this way: if you could add up all of the orbital and rotational motions (or in physics terms, the angular momenta) of all of the Solar System's bodies including the Sun, the Solar System would still spin in a certain direction. The invariable plane is the "disk" of this rotation. The "right-hand rule," as illustrated to the right, can be used to determine north for the invariable plane.

Direction of Orbits

So if you look down onto the rotation axis of Venus from the north side of the invariable plane, you will see that the planet rotates clockwise, in the opposite sense—or retrograde—from the counterclockwise motion of the invariable plane. From this perspective, both the gas giant Uranus and the dwarf planet Pluto are also retrograde rotators.

Pretend that your right hand is the size of the Solar System—or, if it's easier for you, pretend the Solar System is the size of your right hand. If you curl your fingers in the direction of the average orbital motion, your extended thumb points to the north.

However, like the decision to demote Pluto from planet to dwarf planet, the IAU's definition of north isn't universally accepted. Some argue that the IAU definition is flawed because what constitutes a planet's north relies on something outside of the planet itself—the average angular momentum of the rest of the Solar System.

An alternate definition of north for a planet applies the right-hand rule to the planet's rotation only, without regard for the invariable plane. By this definition, Venus is upside down, with an obliquity of 177°, rather than rotating backward with obliquity of 3°. Obliquity is the axial tilt, or the angle between the rotational axis and a line perpendicular to the orbital plane. So instead of rotating retrograde, Uranus is simply lying on its side—obliquity of 98°.

In some ways it's a bit arbitrary: is Venus tilted at 3° and rotating backward or tilted at 177° and rotating forward? For the purpose of calculating a planet's motions, either definition works, and you'll often see obliquities and rotation periods tabulated both ways. There are, however, different implications for a planet's history that go with each interpretation.

Early on, Uranus probably rotated similarly to the other gas giants, but billions of years ago a large collision with another massive body likely toppled it over. As for Pluto . . . well, it's hard to say. It's not unusual for Kuiper Belt objects (KBOs)—and the dwarf planet *is* a KBO—to rotate in the retrograde sense. It is very difficult to envision, however, how Venus could have been turned completely upside down. Therefore most planetary scientists agree that Venus is clearly undergoing retrograde rotation: it was not toppled over but is instead truly spinning backward. Recent theoretical models suggest that two early giant impacts (similar to the Earth's Moon-forming impact) were violent enough to slow down the prograde spin of Venus and reverse its rotation without causing its spin axis to be tilted.

The direction of the moon Triton's orbit about Neptune is opposite to the direction of the planet's rotation. The combination of Triton's orientation around Neptune and Neptune's orientation with respect to the Sun exposes Triton to decades-long seasons similar to those experienced by the planet Uranus.

The Neptunian moon Triton is retrograde in a different aspect of its motion: Triton *orbits* in the "wrong" direction! All of the planets of our Solar System orbit the Sun in the prograde direction. Similarly, all other *large* moons have prograde orbits about their planets, that is, in the same direction as their planet's rotation. For these moons, the orbital axis points in the same direction as the planet's rotational axis (right-hand rule again!). For Triton, the orbital axis points opposite to Neptune's rotation axis. It is the only large moon to have a retrograde orbit.

There are plenty of small moons in retrograde orbit about the gas giants (Jupiter, Saturn, Neptune, and Uranus). In fact, it's more common for these *irregular moons* to have retrograde orbits than not. These moons were likely captured rather than formed in place: their orbits are distant, inclined, and often very noncircular (eccentric).

Triton, however, is no irregular moon. Its orbit is relatively close to Neptune and is very nearly circular. Plus, Triton is large. As the largest of Neptune's moons, it is bigger than the dwarf

Triton and a binary companion approach Neptune. The companion escapes but Triton is captured into a retrograde orbit about Neptune.

planet Pluto and nearly three-fourths the size of our own Moon. By way of comparison, Saturn's moon Phoebe—the largest of the irregular moons—is less than a 10th of the size and has only 0.04% of the mass of Triton.

So why, then, does Triton orbit the "wrong" way? Like the irregular moons, Triton's retrograde orbit implies that it did not form in place around Neptune. Due to the large moon's many similarities to KBOs (such as the dwarf planet Pluto), the most likely explanation is that Triton is a captured Kuiper Belt object. One possible scenario envisions Triton as a member of a binary pair of KBOs that pass too close to Neptune. The other member of the binary pair is ejected by Neptune, leaving Triton in a close orbit about the gas giant. Over time, tidal interactions and gas drag circularize the orbit and synchronize Triton's rotation to it. Like Earth's Moon, the same side of Triton always faces its primary (in this case Neptune).

Another consequence of the retrograde orbit is that Triton experiences *tidal deceleration.* Unlike our own Moon, which is slowly receding from the Earth due to *tidal acceleration,* Triton is moving closer to Neptune. As Neptune tugs on Triton's tidal bulge, the moon is slowed in its orbit. This causes the orbit to decay—Triton spirals closer and closer to Neptune. The icy moon may eventually collide with Neptune or possibly be torn apart by Neptune's gravity, perhaps leaving a ring of icy debris similar to Saturn's ring system.

Although retrograde rotation and retrograde orbits are not unusual in the Solar System, it is unusual for bodies as large as Venus and Triton to exhibit these motions. Such behavior is usually seen in small Kuiper Belt objects and the irregular moons of giant planets. Venus is the only planet unequivocally experiencing retrograde rotation, and Triton is the only large moon with a retrograde orbit. You might even say that Venus and Triton are extreme … in a retro sort of way. Groovy, man.

Most Misinterpreted "Artifact"—The Face on Mars

Taken July 25, 1976, this Viking Orbiter 1 image of the Cydonia Mensae region of Mars has been used as evidence for an advanced alien civilization on Mars. What appears to be a face can be seen in the upper center part of the image. The speckled appearance is due to missing data, called bit errors, caused by problems in transmission of the photographic data from Mars to Earth. One such bit error makes up what appears to be a nostril.

We admit it. There is a massive and ongoing cover-up that has lasted more than thirty years. We've been a part of it from the beginning. We were only nine years old at the time but *They* felt it was important to have us involved in 1976 when the Viking Orbiter sent back the first images of Cydonia Mensae. Most people think the face on Mars is a product of an ancient martian civilization. What people aren't aware of, and what *They* don't want you to know is that . . .

Okay, not really. But wild conspiracy "theories" of this sort have attached themselves to the original Viking Orbiter 1 imagery almost from the beginning. There is nothing mysterious about the images, however. The image caption originally supplied by NASA for the

Cydonia Mensae image pointed out the similarity to a face and attributed it to an optical illusion caused by viewing angle, lighting angle, and shadows. As excited by the prospect of alien life as planetary scientists are (detection of life on Mars was one of the goals of the Viking mission, after all!), they wouldn't be able to keep such a finding secret for thirty minutes, much less thirty years.

To use the phrase popularized by the late Carl Sagan: Extraordinary claims require extraordinary evidence. This hasn't stopped the conspiracy/alien-civilization "theorists" from putting forth all sorts of extraordinary claims with only the Viking 1 photo as evidence.

Working scientists, on the other hand, have a much different definition of *theory*. A scientific theory must consistently fit all the observed facts, make predictions about what future observations or experiments will show, and be testable in such a way that the theory may be shown to be incorrect or incomplete if it cannot account for all observations. A theory is definitely not whatever wild idea or hunch can be thought up. The common, nonscientific usage of the term *theory* is what scientists would call a *hypothesis*.

To test the hypothesis that the face on Mars is just a hill, subsequent missions to the Red Planet have made increasingly better measurements of the Cydonia region. In 1998 and again in 2001, NASA's Mars Global Surveyor (MGS) spacecraft snapped detailed pictures of Cydonia and the face, using the Mars Orbiter Camera (MOC) at a resolution more than 25 times greater than the best Viking images. Confirming the light and shadow hypothesis, these images show a

Comparison of the face on Mars from images taken by Viking 1 on July 25, 1976 (left), by MGS/MOC on April 8, 2001 (center), and by MRO/HiRISE on April 5, 2007 (right).

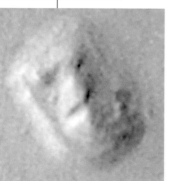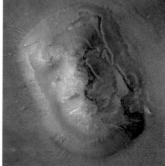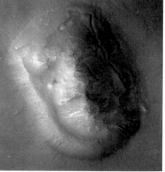

Perspective view of the "face" from the High Resolution Stereo Camera (HRSC) onboard ESA's Mars Express Orbiter.

relatively nondescript hill, much like the buttes and mesas found here on Earth.

In July 2006, the ESA's extremely successful Mars Express Orbiter was finally able to get clear views of the Cydonia region with its High Resolution Stereo Camera (HRSC). Images from multiple passes of the orbiter were combined to produce 3-D perspective views, which once again revealed a small hill showing signs of landslides and other erosional features.

Most recently, NASA's Mars Reconnaissance Orbiter (MRO) took absolutely amazing images of the face mesa in April 2007. The High Resolution Imaging Science Experiment (HiRISE) camera was able to capture images at a resolution of 25 cm per pixel, nearly 180 times better than the original Viking images. At this resolution, objects around 90 cm (35 inches) in size are easily distinguishable. So, once again we are treated to a nice—albeit spectacularly detailed—view of a hill. There are no signs of any structures that were made by anything other than natural geologic processes. The geomorphological hypothesis (it's just a hill!) has survived repeated testing by both NASA and ESA and remains the most widely accepted *scientific* theory of the Cydonian "face."

The Badlands Guardian near Medicine Hat, Alberta, Canada. This geomorphological feature resembles the profile of a Native American in a feathered headdress but was formed via erosion by wind and rain. Although the feature appears to rise above the surroundings, it is actually a depression composed of connecting valleys. An oil well and the access road leading up to it create the appearance of ear-bud-style headphones.

Seeing a face in the shadows of a martian hill is an example of the psychological phenomenon known as *pareidolia,* assigning meaning to vague images or sounds. Just like seeing the shape of an elephant in a passing cloud or religious figures on toasted baked goods, the Mars face is an example of the human mind's great facility for pattern recognition. Some have suggested that our brains are particularly hardwired to detect the human face. In any case, pareidolia, lighting angle, and shadows aside, the detailed HiRISE observations of the Cydonia mesa should lay to rest the hypothesis that an ancient civilization built a humanoid face on the surface of Mars. That hypothesis is no longer consistent with the observations!

If you still think that you can kind of make out the vague hint of what might be a face and don't see how such a feature could possibly be created by natural geologic processes, may we suggest that you take an airplane trip near the city of Medicine Hat in Alberta, Canada. There's a Badlands Guardian there who just might be able to change your mind.

Three different views (ultraviolet, visible, and infrared) of Saturn by the Hubble Space Telescope in March 2003. Notice the hole in the clouds at the south pole associated with a polar vortex.

According to Greek mythology, the giant one-eyed Cyclopes helped determine the fates of the Olympian gods. Born to the ruling gods Uranus (sky) and Gaia (earth), the three Cyclops brothers were skilled blacksmiths—they forged Zeus's lightning bolts, Poseidon's trident, and Hades' helmet of darkness. Uranus feared the powers of the stubborn Cyclopes and banished them underground in the belly of their mother Gaia.

A young Titan named Cronus (also a son of Uranus and Gaia) freed the Cyclopes from imprisonment during his overthrow of Uranus, only to exile them again to the underworld after ascending the throne. Known as Saturn by the Romans, Cronus ruled the ancient world during the bountiful Golden Age.

But a family curse or hex hung over Cronus's head. Like his father Uranus, Cronus was destined to be overthrown by his son. In a desperate attempt to prevent his own ouster, Cronus swallowed each of his

Greek goddess Rhea gives her husband Cronus (Saturn in Roman mythology) a stone in swaddling clothes instead of their son Zeus.

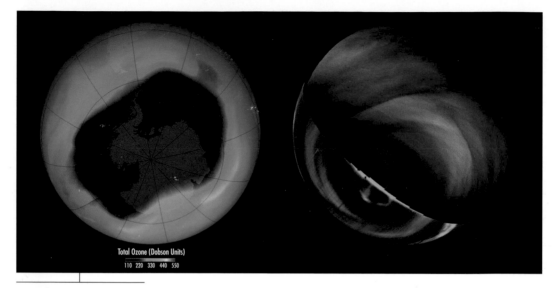

Total Ozone (Dobson Units)
110 220 330 440 550

Polar vortices are found on other planets. The ozone hole over Antarctica (left) delineates the south polar vortex on Earth. Blue indicates areas of ozone depletion, while warm colors show normal levels of ozone. This image from September 24, 2006, shows the largest (in area) ozone hole on record. Venus's south polar vortex (right) as seen from the Venus Express spacecraft in 2007. The ultraviolet portion (blue) depicts cloud structure from the day side at the cloud tops, and the infrared portion (red) shows deeper cloud dynamics on the night side.

newborn children whole. One day, his wife Rhea tricked Cronus into swallowing a stone instead of their son Zeus. Zeus (Jupiter in Roman mythology) returned years later to save his siblings and release the Cyclopes. With the aid of the Cyclopes' thunderbolt, Zeus ultimately defeated Cronus to become the supreme ruler of the cosmos.

A one-eyed monster and a hex still haunt Saturn today, this time in the form of polar vortices.

Polar vortices are no surprise. They likely exist on every planet with a substantial atmosphere. As winds converge poleward from different locations, they begin swirling around the poles. If the planet's rotation is strong enough, north-south winds become sharply deflected in the east-west direction and giant vortices materialize. An intense polar vortex on Earth helps produce the ozone hole over Antarctica by isolating air above the South Pole from the rest of the atmosphere. The polar vortex at Venus's south pole morphs daily from circular to hourglass shaped. Yet Saturn has the most bizarre polar vortices of all.

Like the glare of an angry Cyclops, the southern polar vortex is a bit unnerving. A central eye, surrounded by clouds towering 30–75 km (20–45 miles) high, stares out from the depths at the south pole. Resembling hurricanes on Earth, Saturn's storm has vast spiral arms

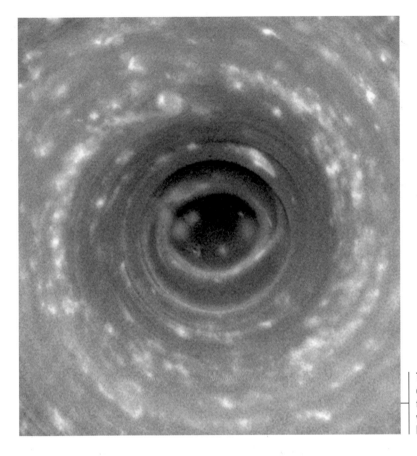

The one-eyed monster polar vortex at Saturn's south pole as observed by the Cassini spacecraft at short infrared wavelengths. Such a storm has never been seen on another planet.

emanating from the central circulation. Fierce winds flow along the eyewall with wind speeds of 160 m/s (350 mph), faster than any on Earth. Since Saturn's eye remains fixed at the pole, this violent storm must get its energy in a different manner than tropical hurricanes on Earth.

The physics of Saturn's eye may be similar to water draining from a kitchen sink. As water spins into the drain, air is also sucked down the center. A strong vortex develops, allowing the sink to drain rapidly (and make a cool slurping sound in the process). It could be that as Saturn's cloudy skies spin around the pole and begin descending into the depths, they drag clear air from above the clouds into the center to produce the spooky, one-eyed vortex.

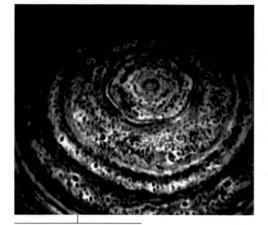

The hexagonal polar vortex at Saturn's north pole as observed by the Cassini spacecraft at infrared wavelengths. Bright areas indicate thick clouds and dark areas show the deep atmosphere.

Meanwhile, the hex hanging over Saturn's head is, in fact, a hexagon . . . actually, multiple concentric hexagons. These hexagons reside near the north pole, and their existence remains an intriguing puzzle. Saturn's hexagonal features are remarkably persistent. They were first observed by Voyager spacecraft in the early 1980s and then again by Cassini in 2006.

Hexagons are commonly found in nature—honeycombs, crystals, and even some cloud patterns on Earth. It's almost as if hexagons are nature's preferred pattern for packing things tightly together. But concentric shapes are different. You can easily place any shape inside itself: circles inside circles, squares inside squares, hexagons inside hexagons. There is no preferential shape or geometry. So why hexagons?

Saturn's hexagons may actually be planetary waves. If you ever have thought that weather seems to come in waves, you were correct. Midlatitude weather systems on Earth are associated with planetary waves that have, on average, three peaks (high-pressure systems) and three troughs (low-pressure systems). To create Saturn's hexagonal weather system, planetary waves would need six peaks to correspond to the vertices of the hexagon. Laboratory experiments and computer simulations have shown that planetary waves will exhibit more peaks if the planet is rotating faster. Saturn rotates once every 10.6 hours, over twice as fast as Earth's rotation rate, so more peaks should be expected.

Like the tragic fate of Cronus, will Saturn's present-day Cyclops and hex destroy the planet? Probably not. But they do cement Saturn's destiny as one of the strangest, and most eye-catching, places in our Solar System.

FrankenMoon—Miranda

When you gaze upon the fractured visage of Miranda, smallest and innermost of Uranus's five largest moons, what is the first thing that you notice? Is it the extremely scarred and varied surface? Like Victor Frankenstein's nameless "dæmon" in the 1818 Mary Shelley novel, the small icy satellite looks as if it was assembled from bits and pieces of other bodies.

When planetary scientists obtained their first close view of Miranda, they didn't know quite what to think. In 1986, Voyager 2 needed to swing very near Uranus to acquire the proper velocity to continue its journey toward Neptune. During the flyby, the spacecraft was able to snap some very close-up shots of Miranda—some of the highest-resolution images of the entire mission.

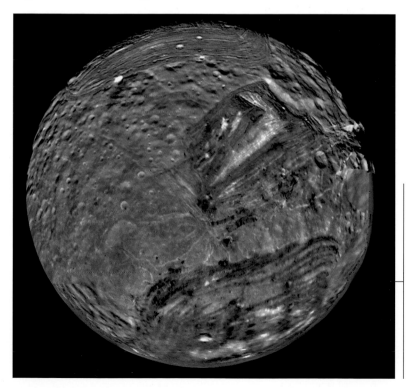

This south polar view of Miranda, the innermost of Uranus's major satellites, shows the patchwork surface of the mysterious moon. The ovoid grooves at the top of the image are part of Elsinore Corona. The ovoid of light and dark bands near the bottom is Arden Corona. The bright "chevron" is part of the trapezoid-shaped Inverness Corona. To the right of Inverness, near the limb, is Verona Rupes, a scarp containing the Solar System's tallest cliff. Interspersed between coronae and rupes are both smooth younger plains and heavily cratered older terrains.

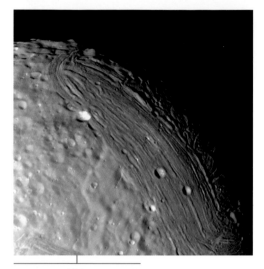

Terrains of different ages can be seen in this Voyager 2 image. Older, cratered, rolling terrain is to the left. The grooved terrain in the center is part of the "racetrack" outer band of Elsinore Corona. On the limb is complex ridged terrain.

The images were baffling. The different geologic regions of the icy moon didn't appear to belong together. Deep scars—the racetrack-shaped coronae—mar Miranda's otherwise relatively smooth face. A bright, icy "chevron" highlights the dark, tortured surface that surrounds it. And curiously, the tallest cliff in the Solar System, perhaps as high as 10 km (6.2 miles) in places, is located on this otherwise inconspicuous moon rather than on a larger planetary body.

How could this patchwork of geologic features wind up on a single tiny, icy body in orbit about Uranus? Such a small, cold moon should not be geologically active. With as much as 80% ice, Miranda has very little in the way of heat-producing elements (decay of radiogenic material is the source of most of Earth's internal heating). And the moon's small size—about 480 km (300 miles) in diameter—means that it would quickly lose any heat that might be left over from its formation.

On the other hand, we do know that impacts are rather common in the Solar System, especially so in the distant past—during the period of Late Heavy Bombardment. Heavy cratering is a feature of pretty much all terrestrial planets and icy moons. Uranus itself is practically lying on its side due to an early large impact. And since Miranda orbits rather close to the ice giant, the tiny moon could easily find itself caught in the cross fire of objects headed for Uranus.

The ubiquity of impacts led to a hypothesis for the origin of Miranda's disjointed surface: the small moon was walloped by a large impact (possibly more than once). The tiny moon must have been totally broken part and then haphazardly reassembled as gravity brought the shattered remains back together. The strange patchwork of surface features is thus a result of the battered moon readjusting to its new stitched-together form.

Although it's a compelling idea, later analyses show that even if such cataclysmic impacts did occur (and they probably did), there would be no evidence of these events remaining on Miranda's surface today. Fortunately, no world shattering is required to explain the patchwork features. As with the animation of Frankenstein's Monster, a jolt of energy is all that is needed to form the features we see today.

We now suspect that the small moon's interior was warmed by a past boost in tidal heating. Theoretical models of the orbital evolution of the Uranian system suggest that Miranda passed through a 3:1 orbital resonance with Umbriel (another of Uranus's larger moons): Miranda traveled around Uranus three times for every one trip made by Umbriel. So, every three orbits Miranda would receive orbit-boosting gravitational tugs from the other moon.

All of this precisely timed, repeated tugging caused Miranda's orbit to grow more eccentric (more egg-shaped) over time. Since Uranus's gravitational pull on Miranda is greater when the small moon is nearer to the planet and diminishes as the moon moves farther away in its egg-shaped orbit, the variation in force caused the small moon's icy surface to be distorted in a fashion similar to the ocean tides of Earth. All of this flexing from tidal tugs could generate significant amounts of heat in Miranda's otherwise cold, dead interior.

With this new (or at least greatly enhanced) energy source, Miranda's interior warmed considerably and caused diapirs of warm ice to rise from below. Extensional tectonics due to blobs of warm, low-density ice pushing up on the surface can produce the faulting, ridges, canyons, and steep cliffs that are observed. A similar process on Earth—with low-density salt instead of warm ice—results in salt domes.

As the orbits of Uranus's moons continued to evolve, however, Miranda was able to escape its resonance with Umbriel. Miranda's orbit would have been quickly brought back to nearly circular—good-bye tidal heating. The existence today of a relatively large orbital inclination (tilt with respect to Uranus's

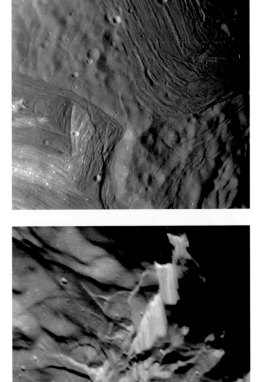

Rolling cratered terrain between Elsinore and Inverness coronae. The scarp Argier Rupes cross-cuts the ridges of Inverness Corona near the "chevron."

The Solar System's tallest cliff, Verona Rupes, can be found on small, icy Miranda. Verona is a sheer cliff face that may be as high as 10 km (6.2 miles)—taller than Mount Everest.

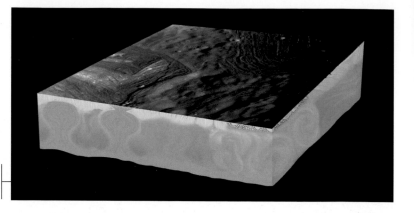

Diapirs of warmer, low-density ice rise from below to create Miranda's observed surface features.

equator)—an order of magnitude greater than any of Uranus's other large moons—is evidence of just such a resonance escape. Bereft of a source of heat, Miranda found itself stuck, frozen into its current jumble of features.

So, while moon-shattering impacts likely occurred, Miranda's FrankenMoon face isn't a result of such events. Instead, a burst of tidal heating in the moon's distant past gave Miranda an energetic jolt. The cold, dead heart of Miranda was warmed and the icy moon was brought to life—geologically speaking—if only temporarily. Now, long frozen and once again inactive, Miranda's scarred visage is merely testament to the dynamic history of Uranus's system of moons.

The spongelike appearance of Saturn's moon Hyperion is probably due to its low total density—half that of water. Incoming objects form deep craters by compressing Hyperion's icy surface rather than by excavating material as on bodies with more typical rocklike densities.

No matter how you look at it, Saturn's moon Hyperion is extremely odd. And you don't have to search very hard to notice the oddities. One of the first things that stands out is the moon's shape—it looks like a potato! The irregular icy moon is roughly 410 × 260 × 220 km (255 × 163 × 137 miles) in diameter along each of its three axes.

Generally, as bodies get bigger, their own gravity tends to pull their

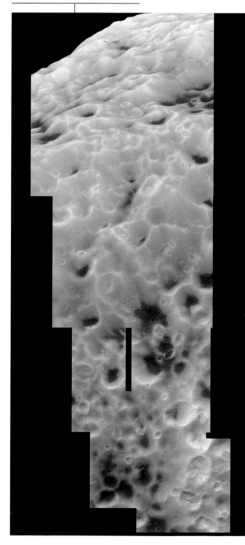

A close-up of Hyperion's craters from September 26, 2005. The Cassini spacecraft approached within roughly 500 km (310 miles) of the moon's surface. Bright crater rims, steep crater walls, and deep, dark crater floors can be seen.

surfaces into a spherical shape. This peculiar moon, however, is one of the largest bodies in the Solar System *not* to have a spherical shape. Neptune's irregular moon Proteus is larger but much closer to being spherical (436 × 416 × 402 km or 271 × 259 × 250 miles) than Hyperion. Hyperion's abnormal shape suggests that it might be a fragment of a bigger moon that was destroyed by a large impact.

Another thing you'll notice with a closer look at Saturn's 16th moon is how pitted and cratered the surface appears. Icy moons are often heavily cratered, especially those that are far enough away from their planet (like Hyperion) to avoid tidal heating. Over time, tidal heating warms the ice and slowly erases surface features. No, the amount of cratering is not what's strange. It's the individual craters themselves.

Hyperion's craters are unusually deep and steep walled. And there is not a lot of ejecta—material blasted out of the crater during impact—that characterizes craters on other planets and moons. It's as if the impacting bodies have punched into Hyperion, compressing the surface ice instead of throwing it explosively outward.

Such compression is possible because of Hyperion's low total density—the small moon would easily float in water. To have such a low density, Hyperion must be composed mostly of water ice. However, the moon's interior is riddled with holes—it's more than 40% empty space inside! Not only does it look like a sponge, but it's porous like a sponge, too.

But the potato shape, the steep-walled craters, and the spongelike porosity are not nearly as odd as Hyperion's bizarre rotation.

On Earth, we're accustomed to a fairly regular rotation rate. The Sun rises in the east and sets in the west, and the mean solar day is 24 hours long. Our days are so regular that they were originally used to define the second (1/86400 of a mean solar day).*

Humans have been predicting the motion of the Sun (and by extension, the periodicities of Earth's orbit and rotation) for centuries: Stonehenge, the Mayan Long Count, the ancient Egyptians. Because where and when the Sun will make its appearance on the horizon each morning is so predictable, we have the phrase "As sure as sunrise." If something is as sure as sunrise, then you know it's definitely going to happen the way you expect it to.

This phrase simply does not apply on Hyperion. Sunrise on the icy potato is just not predictable in any long-term fashion. The days are never the same. Not only does the rotation rate (the length of day) vary erratically, but Hyperion's north pole continually points to a different location in space. Hyperion's rotation is a tumbling chaotic mess.

Hyperion (faint spot at lower left) in orbit about Saturn. This image was taken with the rings edge-on, and shadows of the rings darken the planet's northern hemisphere. Also visible are the brighter moons Tethys and Enceladus in the ring plane.

And we mean "chaotic" in a scientific sense. Even though the equations that predict Hyperion's rotational motion are well known, the solutions to those equations are extremely sensitive. Small uncertainties in the initial position or velocity of the moon (and there are always uncertainties in scientific measurements) can lead to large uncertainties in the predicted motion. Just as meteorologists are unable to predict more than about one week's worth of weather here on our home planet, for Hyperion it is completely impossible to predict the direction of its spin axis after about 300 days—it could be pointed anywhere!

It's not that you can't know where and when the Sun will rise tomorrow on Hyperion, or be able to work out where and when it rose yesterday; but you might not know where it'll be in a few months. And in just under a year, you'll have absolutely no idea. . . . Good luck publishing an almanac! If you're planning a space mission to land on

---

*Due to tidal interactions with the Moon, days are actually getting slightly longer by 1.7 milliseconds every century; "leap seconds" are added every few years to adjust our time clocks accordingly. Oscillations of the cesium atom are now used to define the second.

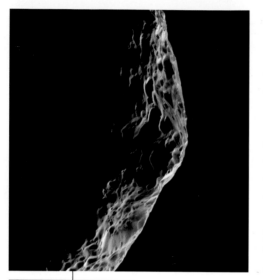

Cassini's parting look at chaotic Hyperion after the spacecraft's close approach on September 26, 2005. In which direction will the raised bump on the surface be pointed when Cassini visits again?

Hyperion and you want to know which face of the moon your spacecraft will be approaching when it gets there, too bad. Hyperion's chaotic rotation will keep you guessing until you get close enough to take some new measurements.

What causes Hyperion to tumble chaotically? Most of the larger moons are tidally locked to their planet—that is, they rotate at the same rate that they orbit. This causes the satellites to keep the same face turned toward their planet at all times; our own Moon orbits Earth in this fashion.

However, the oblong shape of Hyperion prevents such tidal locking. With uneven gravitational torques from both its eccentric orbit about Saturn and a 4:3 orbital resonance with the large moon Titan (Hyperion completes three orbits around Saturn in the same amount of time that Titan completes four orbits), the battered icy lump just can't settle down into a normal orbit with a normal spin. The elongated moon gets twisted and turned so often by Saturn and Titan that it just doesn't know where it's headed next.

So while you may occasionally have days that feel out of control, they are nothing like the chaotic days of Hyperion. Where nothing is "as sure as sunrise" . . . not even sunrise.

The Incredible Shrinking Planet—Mercury

A shrinking planet? We're not making this stuff up, honest! There is evidence on the surface of Mercury that the planet is actually shrinking. And the evidence starts with Mariner 10, the first spacecraft to visit Mercury.

Three separate flybys were made of the innermost planet between 1974 and 1975. One of the first discoveries was a proper estimate for the bulk density of the planet. And Mercury is surprisingly dense for such a small planet.

Mercury is nearly as dense as Earth, a planet whose diameter is about three times larger. Whereas roughly 30% of Earth's density is due to compression—Earth's larger size means that there's more

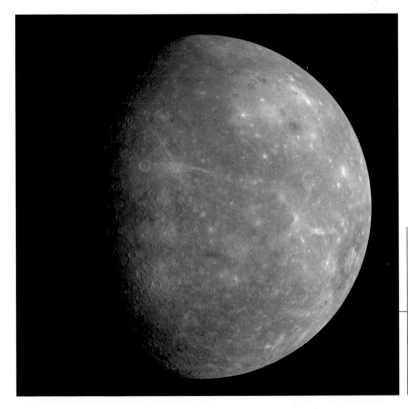

Mercury's previously unseen side was imaged during the MESSENGER spacecraft's January 2008 flyby. Data from three different color filters were combined to produce this false-color view. The Mariner 10 flybys in the mid-1970s revealed large curved cliffs—lobate scarps—that suggest global-scale contraction of the planet's crust. MESSENGER's view of the remaining surface confirmed that the lobate scarps are indeed ubiquitous. Is Mercury shrinking?

"stuff" pressing in on the interior, packing material together more tightly—Mercury is too small to cause such large compression. Its composition must be inherently dense. So, unlike the other rocky planets, silicate rock cannot make up the bulk of tiny Mercury's interior. Most of the planet's mass must be due instead to a large dense core, most likely made of iron.

In addition to a large core, Mercury was found to have a weak—about 1% that of Earth's—magnetic field. Since Earth's dipole magnetic field is thought to be generated by a geodynamo in our planet's liquid outer core, Mercury's magnetic field suggests that the planet's large iron core is still molten (at least partially).

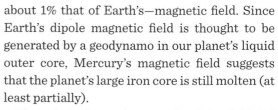

The relatively weak strength of the field led scientists to argue over whether a planet as small as Mercury could retain enough heat inside to keep iron molten. Perhaps the weak present-day field is merely a remnant of a hotter, more active interior from eons past, and the large dense core is no more than a lump of solid iron.

In 2007, however, radar studies of Mercury's rotation showed that slight wobbles in the planet's spin didn't match what we would expect from a completely solid core: Mercury's iron core does indeed have a molten part. The observed weak dipole field then suggests that some form of magnetic dynamo is in operation, probably driven by the cooling and solidification of the molten outer core.

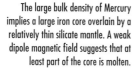

The large bulk density of Mercury implies a large iron core overlain by a relatively thin silicate mantle. A weak dipole magnetic field suggests that at least part of the core is molten.

The primary evidence for *shrinking* of the planet, however, comes from Mariner 10's images. Due to the alignment between the Sun, Mercury, and the spacecraft, Mariner 10 was unfortunately able to image only about 40% of the small planet's surface. But in those areas that were imaged, the spacecraft discovered many large thrust-fault features known as lobate scarps—large, curved cliffs.

The thing is, these cliffs are big—very big! The largest of the known lobate scarps on Mercury—Discovery Rupes—is 2 km (1.2 miles) tall and around 650 km (400 miles) long, greater in overall scale than Earth's Grand Canyon.

Large scarps are not necessarily unique to Mercury. Similarly sized scarps are seen on Mars, such as Amenthes Rupes. So it's not that Mercury's scarps are especially large that makes them stand out (although they *are* large); it's that there are so many of them . . . all over the place. NASA's MESSENGER spacecraft found even more lobate scarps during its 2008 flybys in preparation for a 2011 orbit insertion around Mercury.

There seems to be no concentrated location or preferred direction for the scarps, either. We would expect some form of clustering or directionality if local tectonic processes were shoving blocks of crust together. No, these scarps appear to be a *global* phenomenon that makes the planet resemble a shriveled apple left in the sun for too long. What could possibly be causing this much compression all over the surface of Mercury?

Compressional thrust-faulting forms the lobate scarps. Crust is pushed (compressed) until it breaks (faults) and is thrust up and over neighboring crust.

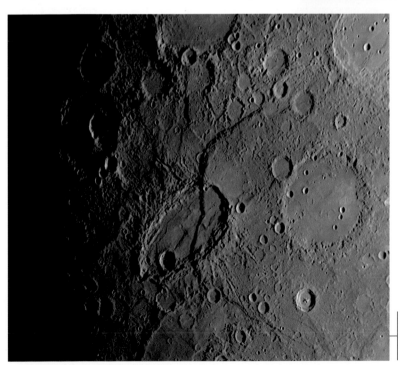

Beagle Rupes, a scarp newly discovered by the MESSENGER spacecraft, cuts across crater Sveinsdóttir on Mercury.

The large fluid core contracts as it cools and freezes into a solid ball of iron. This results in significant shrinkage. Shrinking of the interior causes compression in the outer crust. Thrust faults and lobate scarps form near the surface.

Let's review the evidence. The large density implies a very large iron core—the core radius is about 75% of the size of the entire planet. The rotational behavior suggests that this large iron core is still molten. And from what we know about planetary dynamos, the presence of a magnetic field is indicative of a molten outer core solidifying to form a solid inner core.

Most things contract as they cool, but for molten metals there is an additional shrinkage that typically occurs with solidification; atoms rearrange themselves from a random fluid assemblage into a more closely packed crystalline structure. For iron, this solidification shrinkage can be around 5% of the volume.

So the most likely explanation for the prevalence of lobate scarps on Mercury is that the planet is shrinking as its large iron core cools. Mercury's core makes up such a large fraction of the planet's interior that a shrinkage of a few percent during cooling and solidification would have quite an effect on the surface.

In fact, based on the number and size of the lobate fault scarps observed up to this point, Mercury has probably lost at least 3 km (almost 2 miles) from its radius (currently at 2,440 km or 1,500 miles) simply due the amount of cooling and solidification that has happened so far in the tiny planet's history. Although that seems like a lot of shrinkage, it's nothing compared to the incredible 17 km (11 miles) of contraction estimated for when the entire core has completely solidified. Incredible shrinking planet indeed!

A Perfect Fit—Solar Eclipses on Earth

It's a truly bizarre experience, like someone or *something* suddenly flips a cosmic light switch. Over a span of roughly 40 minutes, the Moon deliberately inches its way across the Sun. The sky dims slowly, almost imperceptibly to the naked eye. And then it happens: day abruptly becomes as dark as night. Stars twinkle in the midday sky, and an eerie halo encircles the dark Moon. After a few minutes of mystical absence, the Sun finally peeks from behind the Moon and daylight returns. The total solar eclipse is over.

Solar eclipses have been messing with people's minds for ages. The ancient Chinese believed that a dragon devoured the Sun; they would bang drums and shoot arrows into the sky to scare the celestial beast. For some cultures, poison drips from the sky during a solar eclipse. Drinking wells are covered in Japan and pregnant Hindu women stay indoors. In 585 BCE, the frightened Middle Eastern armies of Lydia and Medes ended their five-year war immediately following

The total solar eclipse of August 1, 2008, in the Gobi Desert near Jiayuguan, China. This multiple-exposure photograph captures the progression of the eclipse from partial to total to partial again. Bright edges in the sky mark the extent of the Moon's shadow.

Total solar eclipses occur within the umbra, whereas partial eclipses fall within the penumbra.

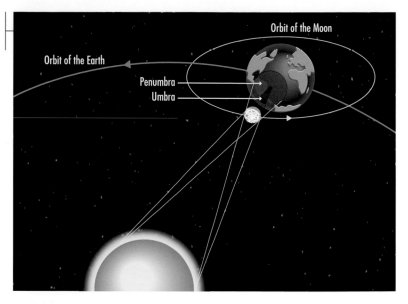

Types of Solar Eclipses in the Solar System

Planet	Number of Moons	Transits	Total Eclipses	Occultations
Mercury	0	0	0	0
Venus	0	0	0	0
Earth	1	0	1	0
Mars	2	2	0	0
Jupiter	63	59	0	4
Saturn	61	51	2*	8
Uranus	27	12	0	15
Neptune	13	6	0	7
Pluto	3	0	0	3
Eris	1	0	0	1

*Not perfect total eclipses because of the moons' irregular shape.

a total solar eclipse. They sealed the treaty with the marriage of a Lydian princess and a Median prince.

Yet there is no *real* ominous threat from a solar eclipse (although if it ends wars . . .). It is merely a rare alignment of the Earth, Moon, and Sun. Two to five times a year, the Moon passes directly between the Earth and Sun. If the alignment is just right—and this happens only for a few minutes every 16 months or so—the Moon completely obscures the disk of the Sun: a total solar eclipse.

What makes a total solar eclipse so rare is the inclination of the Moon's orbit. The Earth-Moon plane and the Sun-Earth plane don't exactly overlap. If the Moon orbited in the same plane as Earth does around the Sun (inclination = 0°), a total solar eclipse would occur every month during the new-Moon phase. But the Moon's inclination of 5.15° is just large enough for the

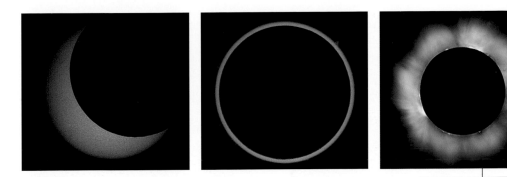

The Moon blocks out a portion of the Sun during a partial eclipse (left). In an annular eclipse (center), the Moon appears just a bit smaller than the Sun. The surface of the Sun is completely hidden during a total eclipse (right), allowing the Sun's diffuse corona to shine radiantly. Earth is the only place in the Solar System to have "perfect" total eclipses—the Moon and Sun have close to the same shape and angular size in the sky.

Moon's shadow to miss Earth during most months. It's only when the new Moon crosses the "zero plane" that we witness a total eclipse of the Sun.

Even when it does occur, a total solar eclipse can't be seen everywhere across the entire planet. Totality happens only within the umbra, or dark shadow of the Moon, which traces out a narrow path roughly 300 km (190 miles) wide on the Earth's surface. A *partial* eclipse occurs in a broader swath of the penumbral shadow. Dedicated eclipse chasers sometimes spend thousands of dollars traveling to remote locations when the path of totality doesn't cross convenient locations. Because the path of totality is so small, a geographic region may see a total eclipse only once every 54 years.

But the rarity of a total solar eclipse on Earth is not what makes it so special. Because the Moon and Sun conveniently appear to be the same size and shape in Earth's sky, the Moon perfectly covers the Sun during a total eclipse. Earth is the only planet in the Solar System with eclipses of such a "perfect fit."

Solar eclipses are actually quite common in the Solar System. They come in three general types: transits, total eclipses, and occultations. Transits happen when the disk of the planet's satellite appears smaller than the disk of the Sun, that is, the ratio of the angular size of the satellite to that of the Sun is less than 1. Occultations occur when the satellite appears larger than the Sun (ratio greater than 1). For total eclipses, the ratio is very close to 1.

Although the Sun is about 400 times bigger than Earth's Moon, it is also 400 times farther away. This makes the two heavenly bodies appear the same size in the sky from the Earth's surface. But the

Solar transits of martian moons Deimos (top) and Phobos (lower sequence) as observed from the surface of Mars by the Mars Exploration Rover Opportunity in March 2004.

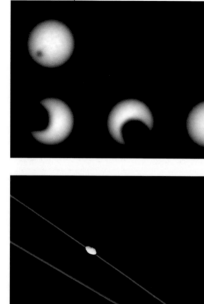

Cassini spacecraft image of Saturn's moons Prometheus (bottom right) and Pandora (top left) near the icy F-ring. Pandora is exterior to the ring and closer to the spacecraft here. The irregular shapes of these moons prevent a "perfect" total solar eclipse.

orbits aren't perfectly circular . . . sometimes the Sun is a bit farther away from Earth, sometimes the Moon is a bit farther away, so the angular sizes change. Because of this, the Moon/Sun ratio varies from 0.92 to 1.06. Some "total" eclipses on Earth are actually transits—*annular* eclipses that show a thin ring of the Sun's surface during maximum coverage. Other total eclipses are borderline occultations, when the Moon appears just slightly larger than the Sun. Yet in every case, the results are totally spectacular.

Only two other satellites in the Solar System produce eclipses with ratios that straddle the magic number 1: Saturn's moons Prometheus (0.74–1.20) and Pandora (0.76–1.07). But these shepherd moons of the faint F-ring look like oblong potatoes. Because of their nonspherical shapes, these moons can never produce perfect total eclipses. If one axis exactly matches the Sun's angular diameter, the other axis will be either too small (part of the Sun will show—a transit) or too large (the Sun's disk is blocked—an occultation).

So it's just the right combination of *size* and *shape* that makes solar eclipses on Earth unlike any other in the Solar System. But it hasn't always been like this, nor will it be like this forever. Because of tidal interactions with Earth, the Moon's average orbital radius is increasing at a rate of 3.8 cm (1.5 inches) per year. The Moon used to be much closer to Earth; it has only been within the last 800 million years that total solar eclipses have even been possible. And since the Moon continues to move farther away from Earth, its angular size decreases slightly each year. In just over 600 million years, Earth will no longer have total solar eclipses at all.

Yes indeed, it is a "perfect" time to be living on planet Earth.

Strangest Life-Form—Humans

Humans are strange. We do really weird things. Stuff that humans consider "normal" is, in the overall scheme of Solar System life, totally off the scale.

Consider for a moment the seemingly lazy task of watching sports on television. The sporting event itself involves complex social interactions—it is planned months in advance, immense stadiums are constructed, fans flock in an organized manner, strict rules of the game are applied and (mostly) followed, scores are tabulated, massive amounts of money are exchanged. All of this activity is conducted through complex written and oral communication, including sophisticated whoops and groans.

Through the incomparable use of technology, the remote sporting

Distinctly human—the 2008 Beijing Summer Olympics had it all: culture, language, technology, art, music, mathematics. It was watched by billions of people around the world. Other life-forms probably would have found the event quite strange, if they could understand it in the first place.

events are broadcast into our living rooms. We *know* that the action isn't happening in front of us. Our brains are able to make the abstract connection that the events are taking place elsewhere. But that doesn't stop us from yelling at the television when our team makes a boneheaded play. And it doesn't keep us from imagining multiple "what if?" scenarios after the game is over.

This may not seem like a big deal to you. After all, watching TV is a passive act, a "no-brainer." But no other species can do all of these things at such a high level—symbolic manipulation, complex language, extensive use of technology, abstract thinking. We make watching sports on television look so effortless because we are so good at it. It's in our genes.

It's not that other species don't have humanlike behaviors. Killer whales have "cultures" with different social structures and lifestyles. Rhesus monkeys have a strong sense of fairness—they will not accept food if doing so will cause another monkey to get an electrical shock. New Caledonian crows manufacture simple tools (plant leaves cut in special ways to produce hooks and barbs) to extract insects from crevices. Elephants can do simple arithmetic. But humans take things to a different level.

The evolution of upright posture and complex thought.

We do have certain physiological advantages over other species. Humans are the only upright, bipedal animal species. Although other primates occasionally stand on two legs to move along tree branches, they use all four limbs for most of their locomotion. By walking upright, humans use only about 25% of the energy of chimpanzees (with the same body weight) walking on all fours. Upright posture also frees our hands to carry food and to develop tools. Perhaps most important, upright posture orients the spinal column to directly support our cranium. Our neck muscles don't have to work as hard as in other animals, which allows humans to have bigger brains than would be otherwise possible.

Now don't get too big-headed. In terms of absolute size, a sperm whale's brain dwarfs the human brain. At ~8 kg (18 lb), the sperm whale's brain is the largest of any animal, modern or extinct. Of course, such a huge animal needs a larger brain to control standard bodily functions. Yet even if you consider relative brain size (the ratio of brain mass to body mass), humans still don't come out on top. Sperm whales have a brain ratio of 0.02% and humans 2% (take that, whale), but the diminutive pocket mouse's brain makes up about 10%. In general, small animals have relatively large brains and large animals have relatively small brains.

Rather than size, what differentiates the human brain is its *complexity*. The human brain contains more neurons in the cerebral cortex (the brain's thick outer shell) than are found in other mammals. Like expensive cables for hi-fi equipment, human neurons also have thicker sheaths of insulation that allow faster transmission with less interference. Along with apes and dolphins, humans have a highly developed cerebellum, a region of the brain that stores patterns of movement and repeated tasks for quick recall—in other words, mental agility.

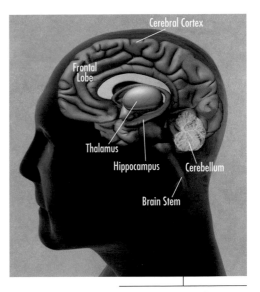

Major parts of the human brain. Planning, ideas, and decisions are processed in the frontal lobes. Memory is stored in the hippocampus, cerebellum, and frontal lobes. These regions tend to be more complex and intricate in humans than in other animals.

Human handiwork from space: The Pyramids of Giza, one of the Seven Wonders of the Ancient World, as photographed from the International Space Station orbiting 350 km (220 miles) above the Earth's surface. The modern-day suburbs of Cairo are encroaching on the ancient lands.

Our free hands and high cerebral activity (at times) allow us to do some pretty amazing things. The rolling wheel. The Pyramids of Giza. The Great Wall of China. Sliced bread. We remember the past and plan for the future. We modify the environment on massive scales that can be detected from space. And although bacteria may have taken an accidental ride aboard a comet, we are the only species to make a *conscious* effort to explore the Solar System.

By reading this book, you are doing something that no other species in the Solar System can do. You think deeply about things well beyond your local cubic meter of space, and you enjoy it! You can analyze the plate motions of the Earth's surface, contemplate the bizarre seasons on Uranus, and ponder the origin of your own existence.

Let's face it: you are a pretty strange life-form.

Supreme Sun

In the grand scheme of things, the Sun may seem like nothing special. After all, the Sun is an ordinary star in the ordinary Milky Way galaxy in an ordinary group of galaxies called the Local Group. It's just one of billions and billions of stars in our universe.

But closer to home, the Sun is a pretty big deal. You might even say that the Sun is the center of attention, the "star," if you will—it *defines* our Solar System. Without the Sun's gravity, the diverse collection of planets, dwarf planets, satellites, asteroids, and comets wouldn't make up a system at all.

And yes, it's hot—a toasty 15 million °C (27 million °F) in the core—definitely the hottest place in our Solar System. As our celestial neighborhood's strongest magnet, this giant ball of fiery plasma generates solar storms that could swallow Earth (if Earth were a bit closer). The Sun's impact is felt over 100 AU away at the heliopause,

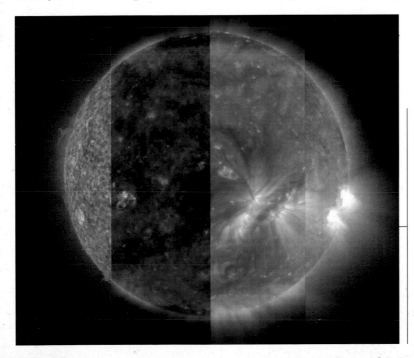

A relatively quiet Sun as observed with four different wavelengths of extreme ultraviolet light by the SOHO spacecraft in November 2005. The red slice (at 30.4 nm wavelength) shows helium at 60,000°C, just above the Sun's surface. Green (at 19.5 nm) indicates ions of iron higher up in the corona at scorching temperatures of 1.0 million °C. Blue (at 17.1 nm) shows iron at even hotter temperatures, 1.5 million °C. On the far right, the upper corona (at 28.6 nm) with temperatures of 2.5 million °C can be seen. High temperatures in the corona are possibly caused by energetic sound waves colliding in the upper corona.

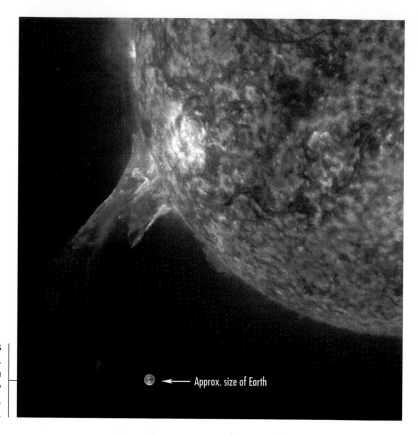

A massive solar prominence erupts from the Sun on July 1, 2002. The relatively cool material of the prominence is trapped by strong solar magnetic fields. Earth is shown for scale.

←— Approx. size of Earth

where the solar wind slams into interstellar particles. Gravitational effects of the Sun dominate at least 100 times farther out, keeping icy bodies within the Oort Cloud from wandering away into outer space.

Even more far-reaching, our local star would appear as a bright beacon of light in the night skies of *other* nearby solar systems in our galaxy. It is the only thing in our Solar System that could easily be detected elsewhere in the universe.

All of these *extra*ordinary characteristics—intense gravity, extreme heat, strong magnetism, powerful winds, and brilliant electromagnetic radiation—ultimately originate from the sheer mass of the Sun. Over 99.8% of the Solar System's mass is condensed into less than one-trillionth of the heliosphere's volume. That's like your car being compressed to the size of the dot in the letter i!

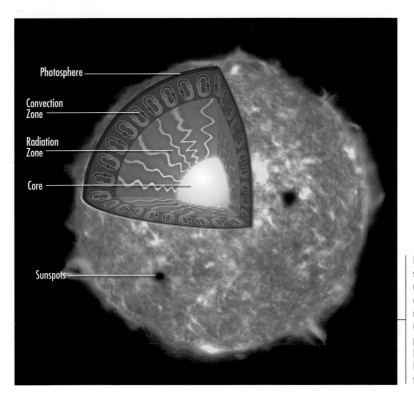

Photosphere

Convection
Zone

Radiation
Zone

Core

Sunspots

Hydrogen fuses into helium in the superheated core (temperature ~15 million °C). X-ray photons transfer energy in the radiation zone (~4 million °C). In the convection zone (~800,000°C), mass movement of plasma transports heat to the surface. Radiation takes over again as light from the photosphere (~5,500°C) travels out into space.

With such a strong concentration of mass, the physics becomes pretty extreme. Intense pressure in the Sun's core (caused by all that weight from above) forces hydrogen atoms uncomfortably close together, so close that new isotopes are forged. The net result: four hydrogen nuclei (protons) fuse together to form a helium nucleus plus two positrons and two neutrinos. Taken together, these new particles have 0.7% less mass than the original four protons. This missing mass is converted into energy—a lot of energy, according to Einstein's famous equation $E = mc^2$.

All of this energy from the Sun's thermonuclear furnace wants to escape. In the dense radiation zone just outside the core, the Sun's plasma is so tightly packed that it can barely move. Energy must be transferred by electromagnetic radiation, but it doesn't happen very quickly. Photons of radiation can't travel very far in this dense layer before being absorbed (and then reemitted in a different direction) by a plasma ion. It takes over 170,000 years for energy to journey through the radiation zone.

Things speed up a bit in the convection zone at a distance 0.7 times the radius of the Sun. The density decreases enough for the plasma to start moving. Hot blobs of roiling plasma are able to rise from the bottom to the top of the convection zone in just over a week! The convection zone also rotates at a faster rate than the rest of the Sun's interior. Because this turbulent rotating plasma is ionized, potent magnetic fields are generated that wreak havoc on the Sun's surface— and elsewhere in the Solar System.

Eventually (sometime in the next 6–7 billion years), the hydrogen fuel within the Sun's core will be exhausted. That's when things get really interesting. Without heat from thermonuclear reactions to support it, the core will collapse. Hydrogen in a compressed shell surrounding the core will then ignite and start to burn. This new energy source will cause the Sun to swell to 250 times its current size and shine 3,000 times brighter!

Our expanding Sun—now a red giant—will easily engulf Mercury and Venus. Bathed in new warmth, Jupiter's icy moons may become "habitable" with liquid water on their surfaces. With its organic soup of hydrocarbons and a red giant incubator, Saturn's moon Titan could possibly develop complex forms of life—the Solar System's new "Earth."

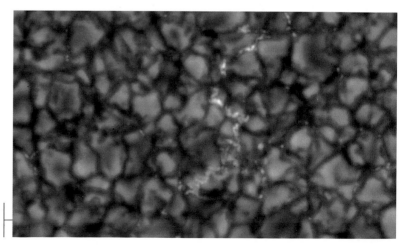

Solar granulation indicates intense convection on the Sun. The largest cells are about the size of Texas.

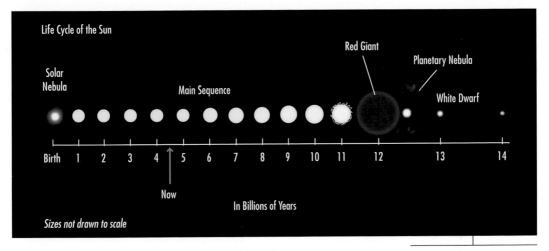

Life Cycle of the Sun

Solar Nebula

Main Sequence

Red Giant

Planetary Nebula

White Dwarf

Birth 1 2 3 4 5 6 7 8 9 10 11 12 13 14

Now

In Billions of Years

Sizes not drawn to scale

Evolution of the Sun from normal yellow star to red giant to white dwarf. The diameter of the red giant will be 250 times larger than that of today's Sun, while the white dwarf will be the size of Earth.

The Sun's nuclear metamorphosis isn't done. When the core temperature reaches 100 million °C, helium will begin to fuse into carbon. This stage induces a severe case of the hiccups in the "star" of the Solar System. The Sun repeatedly shrinks and swells in size (to a maximum radius of 1.2 AU—yes, that's beyond the current orbit of Earth) as helium ignites in multiple shells. With each brilliant helium flash, the pulsing star belches out its hot outer envelopes into the Solar System and produces a dazzling planetary nebula. After less than 400,000 years, only the white-hot carbon core remains, an Earth-sized white dwarf that will slowly cool for eternity.

The ultimate fate of Earth is uncertain. Since the Sun will have lost 25% of its mass during the 12 billion years prior to the red giant phase, the gravitational pull on our planet will be dramatically reduced. Earth could move outward near the current orbit of Mars and thereby escape the red-hot surface of the Sun. Or tidal interactions with the red giant could slow Earth in its orbit, causing our home planet to spiral into the mammoth star toward a searing death.

But for now (and luckily for us), our Sun is in its normal stage, not quite halfway through its "main sequence" life cycle. Yet even in its normal phase, this glowing nuclear reactor of hot convecting magnetism is, well ... *extreme*. There is little doubt that, at least in our Solar System, our ordinary yellow star reigns supreme.

Giant Jupiter

This montage of Jupiter and its moon Io was taken by the New Horizons spacecraft in 2007 on its way to Pluto. The infrared image of Jupiter shows multiple levels in the atmosphere: blue indicates high-altitude clouds and hazes, and red denotes deeper clouds. The Great Red Spot is bluish white in this image. The image of Io is a near-true-color composite that shows a major eruption by the Tvashtar Volcano on the night side. Lava glows red while the bluish plume is illuminated by sunlight.

When it comes to the planets in our Solar System, everything is bigger on Jupiter.* Storms? Almost three Earths could fit inside Jupiter's Great Red Spot, a violent whirlpool that has churned in the Jovian atmosphere for perhaps more than three centuries. Satellites? Not only does Jupiter have four of the six largest moons in the Solar System (including Ganymede, the Solar System's largest moon—bigger than the planet Mercury), the giant planet also has the *most* confirmed satellites with 63. The Jovian system is like a mini

---

*Well, almost everything. There are a few things on other planets that are bigger . . . and we'll leave that short but impressive list for you to ponder.

King of the Planets with the four Galilean satellites (background to foreground): Io, Europa, Ganymede, and Callisto.

solar system doing its own thing 780 million km (480 million miles) from the Sun. Magnetism? Jupiter's magnetosphere is perhaps the single biggest thing inside the Solar System, extending well beyond the orbit of Saturn.

But if you're talking big, mass is what really matters. And Jupiter *really* matters. The gas giant contains over twice the mass of all the other bodies that orbit the Sun—planets, satellites, asteroids, comets, Kuiper Belt objects, and Oort Cloud bodies—combined! Although not quite dense enough to become a fusion-fueled star, the huge blob of hydrogen and helium that is our fifth planet still manages to create some pretty extreme things.

Jupiter's burly mass (and hence flexing gravitational pull) produces the most volcanically active place in the Solar System: hot, smelly Io. Sulfur ions and electrons erupted from Io's interior become rapidly accelerated by Jupiter's intense magnetic field. These near-light-speed particles spawn the harshest radiation zone found among the

Jupiter's colorful palette of dark belts, bright zones, and turbulent eddies as viewed by the New Horizons spacecraft in 2007.

planets. Meanwhile, the massive planet's same penchant for tidal flexing generates enough heat in Europa's interior to create the deepest ocean in the Solar System—a promising petri dish for exotic life-forms lurking in the dark.

Jupiter's highly compressed mass packs a lot of heat within the planet's interior, heat that drives turbulent weather in the atmosphere. Through gravitational contraction, about 70% more heat comes from within Jupiter's interior than reaches the planet from the Sun! This makes Jupiter a very dynamic place—the atmosphere is continually swirling and whirling (in alternating belts and zones that wrap around the planet) in an attempt to get this heat out. As a result, psychedelic clouds of water, ammonia, and methane paint the stormy planet in colorful swaths. It is one of the most beautiful sights in the Solar System.

And it's Jupiter's hurtling mass, by Jove, that dominates the orbital dynamics of the outer Solar System. Jupiter spins faster than any other planet: a day on the giant planet is just under 10 hours long. In its orbit around the Sun, Jupiter propels through space at a whopping 50,000 km/hr (31,000 mph). Taken together, the *rotational* and *orbital* angular momenta of Jupiter account for 99% of the total angular momentum of the Solar System! It's like having a huge ball on the end of a long string that wallops anything and everything in its path as you spin around.

This orbital influence has played a key role in

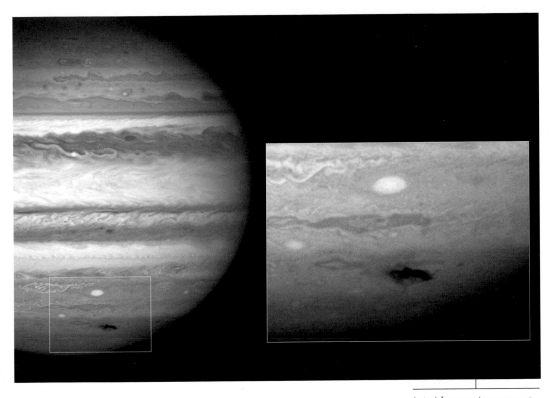

Jupiter's large mass intercepts comets and asteroids that might otherwise hit Earth. Hubble scientists interrupted the calibration of the newly refurbished space telescope to capture this stunning image on July 23, 2009. An impactor, perhaps only a few hundred meters in diameter, left a scar twice the size of the United States.

the evolution of the Solar System. Because of resonances with Jupiter, planetesimals located in orbit between Mars and Jupiter were unable to coalesce into a major planet. The result is the main asteroid belt containing the dwarf planet Ceres and more than a million other smaller asteroids. The period of Late Heavy Bombardment, between 3.8 and 4.1 billion years ago when impact cratering reached a maximum, may also have been caused by orbital resonances with Jupiter.

You might even say that Jupiter has a major "impact" here on Earth. The giant planet has a dual role as both comet shield and asteroid slinger. Comets from the outer Solar System are strongly influenced by Jupiter's gravity. Many of them—like Comet Shoemaker-Levy 9, which slammed into Jupiter in 1994—never make it past Jupiter into the inner Solar System. Complex life has been able to evolve on our home planet because of the relative absence of cometary impacts. On the other hand, Jupiter's persistent gravitational tug can cause asteroids to careen toward the Sun in more elliptical orbits. If such an asteroid crosses Earth's path, the results can be devastating. The K-T impact that wiped out the dinosaurs 65 million years ago could have been nudged into its ill-fated trajectory by the not-so-gentle giant.

For a moment, just try to imagine our Solar System *without* Jupiter. Mars might have ended up larger in size (planetesimals would have been attracted to Mars rather than ejected by Jupiter). With a corresponding denser atmosphere and enhanced greenhouse effect, Mars would be much more temperate and Earth-like. The asteroid belt would be gone, replaced instead by a large rocky planet. The orbits of all the planets would be dramatically altered to compensate for the vast angular momentum currently harnessed by Jupiter. Earth might no longer be in the habitable zone!

Yes indeed, Jupiter is a *giant* in the planetary field, a truly influential player. Our Solar System would not—no, *could* not—be the same without it. That makes Jupiter extreme in a big way.

Sexy Saturn

This fantastic, natural-color mosaic of images taken by the Cassini spacecraft in October 2004 has it all. Pastel-colored bands of clouds show atmospheric dynamics and zonal winds. Ring shadows projected onto the relatively cloudless pale blue north polar region show both that the rings are thick enough to block light and that the gaps are empty enough to let light through. And then, of course, there are the rings themselves. The larger and brighter A- and B-rings are split by the Cassini Division. The narrower Encke Gap, kept open by the tiny moon Pan, can be seen near the outer edge of the A-ring.

Just look at this image of Saturn. There is no way that you can convince us that this planet—the sixth from the Sun—isn't the sexiest thing in the Solar System. Of course, we mean "sexy" in a techno-science-geek way, but come on . . . just look at it! Your eyes *do* keep drifting back to the picture, don't they? Ours too. This is one sweet image of the Solar System's sexiest planet.

Despite the apparent dominance of Jupiter in the Solar System, there are still many ways (above and beyond the spectacular rings) in which Saturn can be considered extreme. For one, the second largest gas giant is the *least* dense of all the planets. In fact, Saturn is less dense than many of its own icy moons! With a bulk density of slightly less than 0.7 g/cm^3 (compare to the density of water at 1 g/cm^3), Saturn is the only planet in our Solar System that would float in your bathtub . . . assuming, of course, that your bathtub is a couple hundred thousand kilometers across.

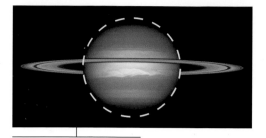

A reference circle illustrates
Saturn's extreme flattening.

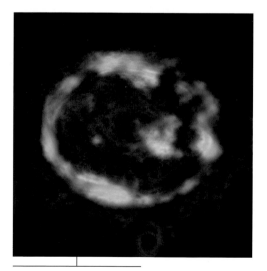

Saturn's north polar hexagon bathed
in auroral light. This composite
infrared image from the Cassini
spacecraft shows both the mysterious
cloud pattern around the north pole
as well as the highly energetic and
variable infrared aurora.

Then there's the bulge. Rapidly rotating planets aren't exactly spherical. The centrifugal force of rotation causes them to bulge outward around the middle. Earth, a moderately fast rotator (once every 24 hours) with fairly rigid and rocky outer layers, has an oblateness, or degree of flattening, of only 0.3%. With a rotation rate greater than twice that of Earth, gas giant Jupiter is flattened by slightly more than 6%. Saturn, however, with its fast rotation (once every 10.5 hours) and low density, experiences nearly 10% flattening—it is by far the most squashed of all the planets.

Because Saturn's rings are so distracting, we often overlook the fact that Saturn has some of the fiercest weather in the Solar System. Massive electrical storms, howling jet-stream winds, and two very odd polar weather patterns (an anticyclonic hurricane with a well-defined eyewall near the south pole, and a hexagonal pattern around the warm north polar vortex) make Saturn a front-runner for weather that is both mysterious and dangerous. And as everyone knows, mysterious and dangerous are sexy . . . even in planetary science.

And when it comes to extreme satellites, Saturn is second to none. The only co-orbital moons in the Solar System—moons that orbit at very nearly the same distances from their parent body—are to be found around Saturn. Both Tethys and Dione have Trojan moons: Telesto and Calypso with Tethys, Helene and Polydeuces with Dione. The Trojans orbit at the same distance from Saturn as their primary moons but 60° ahead and behind at the L4 and L5 Lagrange points, where gravitational forces on the moons and their orbital motions balance.

Even more unusual are the moons Janus and Epimetheus. These moons really *do* share very nearly the same orbit. The 50-km (31-mile) distance between their mean orbits is less than the size of either of

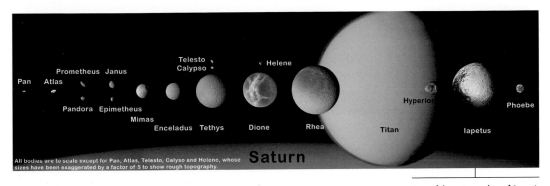

All bodies are to scale except for Pan, Atlas, Telesto, Calyso and Helene, whose sizes have been exaggerated by a factor of 5 to show rough topography.

Saturn

Some of the main members of Saturn's extreme satellite system.

the two moons. A collision would seem to be inevitable. However, in a unique gravitational dance, the two moons swap places every four years in such a way that they are never in danger of touching.

Thanks to the enormous ring created by small Phoebe, the bright-dark, yin-yang surface of Iapetus has the highest surface contrast of any moon, while Mima's impressive Herschel Crater (it looks like the Death Star from *Star Wars,* doesn't it?) hints that the small moon

Saturn

Phoebe

Titan

Iapetus

Saturn's Largest Ring

The vast, tilted, and retrograde-orbiting ring of diffuse ice and dust being blasted off Phoebe may be the source of the yin-yang color variation of Iapetus.

This natural-color mosaic covers approximately 62,000 km (39,000 miles) along the ring plane and reveals the gaps, gravitational resonances, wave patterns, and delicate color variations found throughout the ring system.

barely escaped destruction from the force of a massive impact. The tiny shepherd moons, Prometheus and Pandora, can't resist leaving gravitational graffiti in their wakes (fancy kinks, spirals, and density waves) as they keep the thin F-ring in line. And irregular moon Hyperion chaotically tumbles in its orbit around Saturn, never repeating the exact same rotation.

Don't forget about exotic Titan, the second largest satellite in the Solar System and the only moon with an appreciable atmosphere. It is, in fact, thicker (and hazier) than Earth's own atmosphere. Titan is also the only other body besides Earth with free-flowing liquid on its surface—although this liquid consists of lakes of methane and ethane (the two primary components of natural gas here on Earth) rather than water. So please, no lakeside campfires on Titan!

But for now, the biggest explosions occur on ring-maker Enceladus. Repeated tidal tugging causes cryovolcanic geysers of water and ice to blast forth from the icy moon's "Tiger Stripes." The jets and plumes are the source of water vapor and fine ice crystals that continually replenish Saturn's outermost E-ring.

Oh, and have we mentioned the rings yet? Take a look at them again. Pretty spectacular, right? As the saying goes: if you've got it, flaunt it. Saturn certainly does flaunt those rings. No other planet in our Solar System (or outside our Solar System, that we know of) has such a magnificent ring system. The colors, the structures, the complex dynamics . . . what's not to like?

Another natural-color mosaic from Cassini images, this time taken in July 2008. Notice the changes that have occurred in Saturn during the nearly four years between this image and the one at the beginning of the chapter. Saturn's changing seasons mean that ring shadows no longer drape over the northern hemisphere. Also, the cap of blue previously seen in the northern hemisphere has evolved into zonal color bands. Can you spot the six moons also present in this panorama: Titan, Janus, Mimas, Pandora, Epimetheus, and Enceladus?

So yes, with one of the largest retinues of followers (61 confirmed moons), an understated sense of color (pastel bands of zonal winds and fading blue northern hemisphere), an air of mystery (hexagonal clouds) and danger (sizzling lightning storms), and the ability to really accessorize (spectacular rings), Saturn must be the sexiest planet around.

Imagine for a moment that you are an alien on a long tour of the Solar System. You've hitched a ride on a comet from the distant Oort Cloud as it speeds toward the Sun. Along the way, you see small icy bodies in the Kuiper Belt not much different from your comet. You encounter four giant outer planets and are suitably impressed by Neptune's fierce winds, Uranus's funny tilt, Saturn's amazing rings, and Jupiter's exquisite colors. You pass through the main asteroid belt, easily avoiding a collision with remnants of a planet that never was. With its huge volcanic bulge and dry deep canyons, the red planet Mars hints at a more active past. As you zip around the Sun and head back toward the outer Solar System, you visit the two warmest planets, heavily cratered Mercury and cloud-enshrouded Venus.

And then you see it—a dazzling mélange of deep blue, vivid green, and wispy white. This alluring planet buzzes with activity, from daily earthquakes to fierce electrical storms to complex radio transmissions. There is nothing quite like it in the Solar System.

This space shuttle image of the Strait of Gibraltar and Mediterranean Sea reveals several distinctive features of Earth. The Mediterranean Sea is located on a tectonic plate boundary where the Eurasian (left) and African (right) plates collide. Water can be seen in different phases: liquid water ocean, liquid and ice clouds, and snow in the mountains of Spain (water vapor is invisible). The thin blue line of Earth's atmosphere can be seen near the limb of the planet. Dark green areas indicate vegetation supported by photosynthesis.

Planet Earth is a very strange place indeed. Nowhere else can you find daylight saving time, Ben and Jerry's Chunky Monkey ice cream, or a book on the most extreme places in our Solar System. Earth harbors extreme natural hazards including volcanoes, earthquakes, tsunamis, El Niño, hurricanes, tornadoes, floods, and

"Home Extreme Home." This iconic view of Earth by Apollo 17 astronauts showcases an extraordinary planet: large continents, blue surface water, frosty ice caps, dynamic weather systems, and signs of life.

Weird stuff in the Solar System found only on Earth: bee pollination, soccer games, burnt toast on a clothesline, Hawaiian sunsets.

droughts. Its large moon—one of the two largest known moons in the Solar System relative to the size of the parent planet*—influences seemingly disparate phenomena such as ocean tides, animal migration, and romantic interludes. Yet what truly distinguishes Earth from other places in the Solar System involves things that we take for granted every day: a moving, changing land; water in three phases; an oxygen-rich atmosphere; and life.

Although tectonic processes are evident on other planetary bodies such as Venus and Europa, only Earth's surface moves and shifts on a global scale. Fifteen major tectonic plates float atop the upper mantle's ductile asthenosphere. It's not a quiet, gentle act: giant mountain ranges thrust upward, earthquakes transform the landscape, volcanoes violently create new land, and old crust dives deep into the mantle. This continual recycling allows most of

---

*Pluto's moon Charon is larger in relative terms. Since the center of mass of the Pluto-Charon system lies outside of Pluto's surface, Pluto-Charon should probably be considered a binary system rather than a planet-moon system.

The Earth system consists of strongly interacting components of air, water, land, and life. These components work together to make Earth unique.

the carbon on Earth to reside in the lithosphere (crust and upper mantle) rather than in the atmosphere as on Venus. Many scientists think that global plate tectonics could not exist without a liquid water ocean to help lubricate the sliding plate motion at subduction boundaries.

Water is a truly amazing substance. It stores more heat per mass than any other compound. Swimming pools remain relatively cool in the summer because water stores a lot of heat without dramatically raising its temperature. Like a giant swimming pool, the ocean regulates temperature through massive heat storage and strongly influences Earth's weather and climate.

Contrary to popular belief, water is also extremely toxic (hence the petitions to ban dihydrogen monoxide). Two slightly positive hydrogen atoms attached to a slightly negative oxygen atom help water to dissolve obstinate substances, from majestic mountains to sticky food on a baby's bib.

Earth is the only place in the Solar System where water exists on the surface in all three phases: gas, liquid, and solid. Venus is too hot, Mars is too cold, but Earth's temperature is just right for water (the Goldilocks Principle of planetary science). Over 70% of the Earth's

No other place in the Solar System glows at night like Earth. This composite image by the Defense Meteorological Satellite Program (DMSP) Operational Linescan System (OLS) shows *artificial light illuminating the globe.*

surface is covered in liquid water and 12% in snow and ice, thus making Earth "the Water Planet" or "the Blue Planet."

Yet Earth could also be called "the Green Planet." Mars may have supported microbial life in the distant past, Europa may have extreme life-forms deep in its liquid ocean, and Titan may have the amino acid building blocks for life, but complex life thrives only on one planet: Earth.

We don't know exactly how many different species live on Earth. Roughly 1.6 million species have been identified (mostly insects), but estimates for the total number of species range from 30 to 100 million. One of these species in particular has modified its environment so significantly that Earth as viewed from space looks much different than it would have a century ago—now including artificial lights, aircraft contrails, and shrinking rainforests.

Life adds another curious characteristic to the mix—an oxygen-rich atmosphere. Roughly 21% of Earth's atmosphere is made of molecular oxygen, a by-product of photosynthesis and used by animals to convert food into energy. No other planet has so much oxygen. It's actually a bit of a precarious and fragile situation. An amazingly volatile substance, oxygen stokes the combustion process and makes Earth a dangerous place to live. In addition, molecular oxygen O_2 and atomic oxygen O

can combine to form ozone O_3. Located 25 km (15 miles) above the surface, the ozone layer absorbs harmful ultraviolet radiation and prevents it from reaching the surface. Without the ozone layer, life as we know it would not exist on Earth.

These unusual features of Earth are so commonplace, so *normal* to us, that we often lose sight of how truly extraordinary our planet is. And these unique traits do not exist in isolation. Like the human body with its various systems (cardiovascular, respiratory, skeletal, nervous, etc.), Earth consists of interacting components that support one another. The whole is greater than the sum of its parts—land, water, atmosphere, and life—synergistic ingredients working together to form the complex, delicately balanced system that we call home. The result: perhaps the most remarkable and extreme place in the Solar System.

Glossary

accretion: growth of an object by gravitational attraction and capture of additional matter

amino acid: organic molecule that serves as a building block for proteins and also aids in metabolism

angular momentum: quantity describing the rotational state of a system that depends on the rotational inertia (resistance to rotational motion) and angular velocity (rate of change of the angle); angular momentum is conserved unless acted on by an external torque

antipodes: point on the surface of a body that is diametrically opposite to a given location such that a straight line between the two points passes through the center of the body

aphelion: point in a body's orbit where it is farthest from the Sun

Archaea: major domain of life consisting of single-cell organisms, genetically different from bacteria, that often live in extreme conditions

asteroid: Small Solar System Body in orbit about the Sun that is smaller than a dwarf planet but larger than a meteoroid and that does not have a visible coma when near the Sun

asthenosphere: mechanically weak, low-viscosity region of the upper mantle of the Earth below the lithosphere

astrobiology: the study of the origin, evolution, distribution, and future of life in the universe

astronomical unit (AU): average distance between the Earth and the Sun—approximately 150 million km (93 million miles)

auroral footprint: location in the polar regions of Jupiter where electric currents from the moons Io, Europa, or Ganymede produce intense auroral displays

Bacteria: major domain of life consisting of single-cell organisms

basalt: volcanic rock relatively rich in oxides of magnesium (MgO) and calcium (CaO) but low in silica (SiO_2)

binary system: pair of bodies that orbit about a mutual center of mass that falls outside the surface of either body

bolide: large crater-forming projectile whose precise nature is not known (e.g., rocky or metallic asteroid or an icy comet)

bow shock: shock caused by the Sun as it moves through the interstellar medium

butte: isolated hill with steep, often vertical sides and a small, relatively flat top (smaller than a mesa)

caldera: large, usually circular depression at the summit of a volcano formed by collapse when magma is withdrawn or erupted from a shallow underground reservoir

carbonaceous: containing carbon

Centaur: Small Solar System Body that orbits the Sun between Jupiter and Neptune and crosses the orbits of one or more of the giant planets

center of mass: point in a system at which all the mass may be considered to be concentrated

chaotic: inherently unpredictable (but not random) due to extreme sensitivity to initial conditions

chemosynthesis: process of producing carbohydrates through chemical reactions without sunlight

chromosphere: thin layer of solar atmosphere between the photosphere and corona; literally "color sphere"

coma: nebulous envelope of gas and dust surrounding the nucleus of a comet

comet: Small Solar System Body in orbit about the Sun that exhibits a visible coma (and sometimes a tail) when near the Sun

condensation: phase change from gas (vapor) to liquid; heat is released into the environment during condensation

continental drift: movement of the Earth's continents relative to one another

convection: heat transfer via blobs of hot material rising and blobs of cool material sinking

co-orbital moons: natural satellites that orbit at the same distance (or nearly so) from their parent planet

corona (geologic): ovoid tectonic feature characterized by annular fractures and commonly with associated radial fractures and lava flows; thought to be produced by a buoyant mantle diapir that flattens and spreads at the base of the lithosphere and causes fracturing, uplift, and melting

corona (solar): outermost, transparent region of the Sun's atmosphere; adjacent to the chromosphere with temperatures exceeding 1 million °C

coronal mass ejection: expulsion of plasma (and magnetic fields) from the Sun's corona, likely produced by unstable magnetic fields in the corona

cosmic radiation: high-energy particles (mostly protons) traveling through space that originate from various sources throughout the universe, including the Sun

Cretaceous period: geological period 145–65 million years ago, generally characterized as very warm and wet; dinosaurs were at their most diverse during this period

cryolava: material erupted from cryovolcanoes, possibly icy slurries of water and other volatiles

cyclone: inward circulation of air masses about a low-pressure system, rotating counterclockwise in the northern hemisphere and clockwise in the southern hemisphere for planets with prograde planetary rotation

density: intensive property of matter that is defined as the ratio of an object's mass to its volume

deposition: phase change directly from gas (vapor) to solid (ice), bypassing the liquid phase

diapir: geologic structure consisting of mobile material forced upward into more brittle surrounding material

differentiation: process of separating out different constituents of a planetary body into compositionally distinct layers

dipole molecule: molecule, such as H_2O, with a negative electrical polarity on one side and a positive electrical polarity on the other side

DNA: complex organic molecules in the shape of a double helix that encode genetic information within cells

domain: top-level branch of the classification of life based on genetic sequences, specifically in ribosomal RNA; the three domains are Bacteria, Archaea, and Eucarya

dust tail: cometary tail of debris left behind in a comet's orbit

dwarf planet: celestial body that (a) is in orbit around the Sun, (b) has sufficient mass for its self-gravity to overcome rigid body forces so that it assumes a hydrostatic equilibrium (nearly round) shape, (c) has not cleared the neighborhood around its orbit, and (d) is not a satellite

eccentricity: measure of the shape of an orbit; values of eccentricity for various shapes: circular orbit (e = 0), elliptical orbit (0 < e < 1), parabolic orbit (e = 1), hyperbolic orbit (e > 1)

ejecta: debris that is ejected during the formation of an impact crater

elastic solid: solid material capable of returning to its initial form or state after a deformation-inducing stress is removed

energy: ability of a system to do work

equator: intersection of a planet's surface with the plane perpendicular to its axis of rotation and containing the center of mass; the imaginary line dividing the northern and southern hemispheres

equinox: time of year when the Sun is directly over the equator; days and nights are of equal length

erosion: physical removal of solids (sediment, soil, rock, etc.) usually due to transport by wind, water, or ice, or by downslope creep under the force of gravity

ethane: simple hydrocarbon compound C_2H_6

Eucarya: major domain of life consisting of organisms with cell nuclei, including plants, animals, fungi, and protozoa

evaporation: phase change from liquid to gas (vapor); heat is removed from the environment during evaporation

exobiology: the study of life beyond Earth; alternate name for astrobiology

extremophile: organism that thrives in and maybe requires (physically or chemically) extreme conditions that are hazardous to most life as we know it

eye: central clear portion of a tropical cyclone

eyewall: innermost circular band of clouds surrounding the eye of a tropical cyclone

Fermi glow: glowing particles created by the bow shock in the solar wind

filament: dark linear feature on the Sun's disk produced by viewing a solar prominence edge-on with the bright solar photosphere in the background

frost line: particular distance from the Sun in the forming Solar System where temperatures are sufficiently low for hydrogen compounds such as water, ammonia, and methane to condense into solid ice grains

full Moon: location in the Moon's orbit when the Moon is on the opposite side of Earth from the Sun

Galilean satellites: four largest moons of Jupiter (Ganymede, Callisto, Io, and Europa), discovered by Galileo

gamma ray: region of the electromagnetic spectrum less than 0.01 nm in wavelength, shorter than X-rays

gas giant: large planet composed primarily of gas instead of rock or other solid matter

gas tail: cometary tail of ionized gas vaporized from the nucleus and carried away by the solar wind; it points directly away from the Sun

geodynamo: self-sustaining process responsible for maintaining the Earth's magnetic field in which the kinetic energy of convective motion of the Earth's liquid core is converted into magnetic energy

geyser: intermittent discharge of liquid and vapor ejected turbulently

glycine: smallest amino acid found in proteins

granulation: grainy appearance of the Sun's surface produced by convection

graupel: soft hail formed by supercooled liquid droplets freezing on snow crystals

gravitational contraction: process in which mutual gravitational interactions cause the interior of a body to compress and thus generate heat

habitable zone: region for which the distance from the Sun is suitable for the existence of liquid water on the surface of a planet

hail: precipitation in the form of layered ice pellets or stones with diameters greater than 5 mm (0.2 inches)

heat: transfer of energy associated with a difference in temperature, flowing from higher to lower temperature; often confused with infrared radiation

heliopause: outer surface of the heliosheath, where the heliosphere meets the interstellar medium

heliosheath: outer region of the heliosphere beyond the termination shock where interstellar gas and the solar wind begin to mix

heliosphere: bubble in space being "blown" by the solar wind that envelops the Solar System

hotspot: anomalous region of magmatism or high temperatures, usually isolated by quieter or cooler surrounding areas

hurricane: tropical cyclone in the Atlantic and eastern Pacific oceans

hydrocarbon: organic compound consisting entirely of hydrogen and carbon

hydrological cycle: movement of fluid (water on Earth, methane on Titan) in the atmosphere, on the surface, and underground involving the processes of evaporation, condensation, and precipitation

hydrothermal vent: fissure on the seafloor that releases superheated water into the ocean

hypothesis: tentative explanation for an observation, phenomenon, or scientific problem that can be tested by further investigation; if observation and experimentation show a conclusion to be false, the hypothesis must be false

ice giant: subclass of gas giant planets composed mostly of water, ammonia, and methane with hydrogen and helium predominantly in the outermost regions

ice sheet: mass of glacier ice that covers surrounding terrain and is greater than $50,000 \text{ km}^2$ (20,000 square miles) in area

impact basin: impact crater with a diameter greater than 300 km (190 miles)

inclination: angle between the orbital plane of an object and the plane of the Solar System

induced magnetic field: secondary magnetic field formed by moving a conductor through an external magnetic field

infrared: region of the electromagnetic spectrum from $0.7\mu m$ to $1,000\mu m$ in wavelength, longer than visible wavelengths of light

infrared window: range of infrared wavelengths in which very little radiation is absorbed by the atmosphere

inner planet: planet whose orbit is nearer to the Sun than the asteroid belt (i.e., Mercury, Venus, Earth, and Mars)

interstellar medium: matter (mostly gas and dust) that pervades interstellar space between the stars within a galaxy

invariable plane: plane passing through the center of mass of the Solar System perpendicular to the angular momentum vector

ion: atom or molecule whose total number of electrons does not equal the total number of protons and thus has a net positive or negative electrical charge

ionization: converting an atom or molecule into an ion by adding or removing charged particles

ionosphere: upper atmosphere consisting of charged particles ionized by solar radiation

iridium: chemical element with atomic number 77; rare on Earth but enriched in meteorites

irregular moon: nonspherical moon with one axis significantly longer than the other axes

irregular moon (orbital): natural satellite that has an inclined, distant orbit that is also often eccentric and retrograde

isotope: atom of a chemical element with the same number of protons but different number of neutrons

jet: narrow current of high wind, often in the zonal (east-west) direction

jet stream: jet of wind aloft produced by a strong temperature gradient at lower levels; planetary-scale jet streams often determine the path of large weather systems

Jovian: of, relating to, or similar to the planet Jupiter

Jurassic period: geologic period 199–145 million years ago, generally characterized as warm and humid and dominated by large dinosaurs

katabatic wind: downslope wind caused by gravity; carries high-density cold air from high altitudes to lower altitudes

Kelvin temperature scale: temperature scale whose zero point (0 K) represents the theoretical absence of thermal energy; 0 K = $-273.15°C = -459.67°F$

kinetic energy: energy associated with motion

K-T impact: asteroid impact considered to be the cause of mass extinction 65 million years ago at the Cretaceous-Tertiary geologic boundary

Kuiper Belt: region of the Solar System extending beyond the planets from the orbit of Neptune (approximately 30 AU from the Sun) to roughly 55 AU from the Sun

Lagrange points: locations in space where gravitational forces and the orbital motion of a body balance; there are five such points, denoted L1 through L5

Late Heavy Bombardment: period from 4.1 to 3.8 billion years ago during which the inner planets experienced heavy crater impacting, possibly due to changes in orbits of Jupiter and Saturn

lava: molten rock expelled by a volcano during eruption

leap second: one-second adjustment to the Coordinated Universal Time (UTC) time scale that keeps it close to mean solar time

lithosphere: rigid outer layer of the Earth made up of the crust and upper mantle

lobate scarp: long, sinuous, clifflike feature consisting of a series of connected lobes; found mainly on Mercury and interpreted as compressive thrust faults

Local Group: group of 30 galaxies in our celestial neighborhood, one of which is our own galaxy, the Milky Way

magma: molten rock found beneath the surface of a planet

magma ocean: large region of molten magma likely present on the surfaces of the Earth, the Moon, and perhaps other planetary bodies during the early stages of the Solar System

magnetic braking: slowing of the Sun's rotation due to the outward transfer of angular momentum by solar wind particles

magnetic dipole: magnetic field in which the field is considered to emanate from two opposite poles, as in the north and south poles of a magnet

magnetic dynamo: process through which a rotating, convecting, and electrically conducting fluid acts to maintain a magnetic field

magnetic field: region surrounding a magnet, electric current, or changing electric field in which magnetic forces are observable

magnetite: iron oxide that is the most magnetic naturally occurring mineral on Earth; it's sometimes found in bacteria and animals

magnetosphere: region dominated by a planet's intrinsic magnetic field

main asteroid belt: collection of Small Solar System Bodies whose orbits lie between the orbits of Mars and Jupiter

main sequence: phase of a star's evolution in which hydrogen is fused into helium in the core

mantle: part of the interior structure of a planet located between the outer crust and central core

mantle plume: narrow column of hot material that rises up through a planetary mantle via thermal buoyancy

mare (maria, pl.): large, dark basalt plain on the Moon that resembled a sea to early astronomers; formed by impact or volcanic melting

mass: measure of the amount of matter in an object; or a measure of the strength of the object's interaction with a gravitational field

mechanical strength: material's ability to withstand an applied stress without failure

mesa: elevated area of land with a flat top and sides that are usually steep cliffs

meteor: visible streak of light caused by a meteoroid heating up by friction as it flies through the atmosphere

meteorite: meteoroid that survives travel through the atmosphere and impact on a planetary surface

meteoroid: solid object moving in interplanetary space, of a size considerably smaller than an asteroid and considerably larger than an atom

methane: simple hydrocarbon compound CH_4

micrometer (μm): 10^{-6} meters, about the size of *E. coli* bacteria

micron: alternate name for micrometer

microwave: region of the electromagnetic spectrum from 1 mm to 1 m in wavelength, longer than infrared wavelengths

midocean ridge: underwater mountain range that marks where tectonic plates are gradually moving apart

Milky Way: spiral-shaped galaxy that contains our Solar System as well as billions of other star systems

monsoon: seasonal winds that bring moisture from the ocean to a heated landmass; torrential rains may occur during monsoon season

nanometer (nm): 10^{-9} meters, about the size of a complex molecule chain (such as DNA)

neutrino: tiny elementary particle with neutral charge and such small mass that it can pass through ordinary matter

new Moon: location in the Moon's orbit when the Moon is on the same side of Earth as the Sun

nonperiodic comet: comet with a parabolic or hyperbolic orbit that leaves the Solar System after a close approach to the Sun

nuclear fusion: process of light nuclei (such as hydrogen) combining to form heavier nuclei and releasing large amounts of energy

nucleobase: parts of DNA and RNA that spell out the genetic code and make the rungs of the DNA ladder

obliquity: angle between a planetary body's equatorial plane and its orbital plane, also known as the tilt of the rotational axis

occultation: passage of an apparently larger celestial body across an apparently smaller body; the smaller body is completed obscured

Oort Cloud: immense spherical cloud made up of billions of icy bodies surrounding our Solar System; extends about 30 trillion km (18 trillion miles) from the Sun

opposition: time when a planet, when viewed from Earth, is directly opposite the Sun

orbit insertion: maneuver performed by a spacecraft designed to allow the spacecraft to be captured into orbit around a planet or other body

orbital inclination: angle between the plane of the orbit of a planet and a reference plane (the ecliptic plane for planetary orbits or the equatorial plane of the primary planet for natural satellites)

orbital resonance: *see* resonance

organic material: chemical compounds that contain carbon

outgassing: slow release of volatile gases from the interior of a planetary body

ozone hole: depletion of stratospheric ozone over Antarctica in southern spring facilitated by human-manufactured chlorofluorocarbons (CFCs)

ozone layer: layer in Earth's atmosphere 10–50 km (6–31 miles) above the surface where over 90% of the ozone resides; it absorbs over 90% of the incoming ultraviolet radiation from the Sun

pancake dome: unusual type of volcano found on the planet Venus with a broad, flat profile similar to a shield volcanoe but 10–100 times larger than volcanic domes of Earth

Pangea: supercontinent consisting of all of Earth's landmass that existed roughly 250 million years ago

panspermia: hypothesis that states that life is spread throughout the universe and likely arrived on Earth from outer space

pareidolia: psychological phenomenon involving a vague and random stimulus (often images or sounds) being perceived as significant

penumbra: lighter part of a shadow caused by a partial occultation or eclipse

perihelion: point in a body's orbit where it is nearest to the Sun

periodic comet: comet with a circular or elliptical orbit that repeatedly makes a close approach to the Sun

Permian extinction: mass extinction of 95% of the Earth's species roughly 251 million years ago

photochemistry: study of chemical processes associated with the absorption and emission of light

photon: massless elementary particle that transfers electromagnetic energy

photosphere: Sun's surface that radiates most of the sunlight into space; literally "light sphere"

photosynthesis: process in plants and other organisms that converts sunlight, carbon dioxide, and water into carbohydrates and oxygen

planet: celestial body that (a) is in orbit around the Sun, (b) has sufficient mass for its self-gravity to overcome rigid body forces so that it assumes a hydrostatic equilibrium (nearly round) shape, and (c) has cleared the neighborhood around its orbit

planetary nebula: collection of glowing gas and plasma ejected from a star during its latter stage of evolution

planetary wave: large-scale undulation of pressure and wind in the atmosphere, often apparent as a meander in the jet stream

planetesimal: solid object arising during the accumulation of planets whose internal strength is dominated by self-gravity and whose orbital dynamics are not significantly affected by gas drag; objects larger than approximately 1 km (0.62 mile) in the solar nebula

plasma: fourth state of matter consisting of an electrically conducting, partially ionized gas in which the ions and electrons can move independently (solid, liquid, and gas are the other three states of matter)

plate tectonics: the theory that describes the large-scale motions of Earth's lithosphere, which is broken up into a number of large and small plates that are moving relative to each other

polar cap: region near the poles of a planet or moon that is covered by ice

polar hood: cloud coverage near the poles that obscures the surface below

polarity: in magnetism, the "north" or "south" orientation of a magnetic field

positron: antimatter counterpart of an electron, but with a positive charge

prograde motion: rotational or orbital motion in the same direction of the common direction of other bodies in the system

prograde rotation: rotation about the planet's axis in the same sense as the dominant orbital motion of the Solar System

prominence: large, dense concentration of plasma, often in arcs or loops trapped by magnetic field lines that emerge from the Sun's photosphere into the corona

radar backscatter: energy in a radar pulse that is scattered back toward the source antenna

radiation: heat transfer by emission and absorption of electromagnetic energy

radiation balance: the balance of incoming and outgoing radiation; a surface will tend to warm (cool) with more incoming (outgoing) radiation

radio telescope: radio antenna (typically a large parabolic dish) that collects data in the radio frequency portion (3 Hz–300 GHz) of the electromagnetic spectrum rather than visible light (400–790 THz), as is the case for traditional optical telescopes

radio wave: region of the electromagnetic spectrum greater than 1 m in wavelength, longer than microwaves

radiogenic: of or relating to a product of radioactive decay

reconnection: process in which magnetic field lines with opposite polarity "touch" each other and connect, producing a new magnetic configuration and releasing large amounts of energy

red giant: giant star with a relatively low surface temperature, which produces its red color

residual ice cap: permanent ice cap left over after the seasonal (or temporary) ice cap disappears

resonance: orbital arrangement between two (or more) orbiting bodies such that they exert a regular, periodic gravitational influence on each other, usually due to their orbital periods being related by a ratio of two small integers

resurfacing: renewal of a planetary surface by geologic processes

retrograde motion: rotational or orbital motion in the opposite direction of the common, prograde direction of other bodies in the system

rift valley: linear-shaped lowland between highlands or mountain ranges created by the action of a geologic rift or fault

right-hand rule: method for finding the direction of a rotation vector by curling the fingers of the right hand to match the curvature and direction of motion (the upraised thumb indicates the direction of the rotation vector)

RNA: complex organic molecules, usually single stranded, that are responsible for synthesizing proteins and decoding genetic information within cells

Roche limit: minimum distance at which a moon is able to orbit a planet without being pulled apart by tidal forces

rotational bulge: increase in a planet's oblateness due to the centrifugal force of rotation

rupes: scarps or cliffs on planets other than Earth; Latin for "cliffs"

salt dome: structural dome formed when a thick bed of salt (or other evaporite minerals) at depth intrudes upward into surrounding rock, forming a mushroom-shaped diapir

saturation: condition in which, after a sufficient increase in a causal action, no further increase in the resultant effect is possible

scarp: continuous line of cliffs produced by vertical movement along a fault or by erosion

sea level: average height of the sea surface (measured with respect to a suitable reference)

seafloor spreading: process by which the ocean floor is created and extended when two tectonic plates move apart

seasonal ice cap: temporary ice cap that grows and decays from season to season

shepherd moon: small moon that orbits near the outer edges of rings or within gaps in the rings and maintains a sharply defined edge to the ring via gravitational interactions

shield volcano: large volcano with shallow-sloping sides

shock wave: type of propagating, energy-carrying disturbance characterized by an abrupt, nearly discontinuous change in the characteristics of the medium: an extremely rapid rise in pressure, temperature, and density of the flow

Small Solar System Body: object in the Solar System that is neither a planet nor dwarf planet, including most of the Solar System asteroids, Trans-Neptunian Objects, comets, and other small bodies

solar flare: large explosion of magnetic energy from the Sun produced by reconnection of twisted magnetic field lines emanating from sunspots

solar nebula: rotating cloud of gas and dust from which the Solar System formed

solar prominence: large, bright feature consisting of cooler plasma (similar in composition to the chromosphere) anchored in the photosphere and extending outward from the Sun's surface into the solar corona

solar wind: stream of plasma flowing outward from the upper atmosphere of the Sun

solstice: time of year when the Sun makes its furthest passage north or south of the equator

spectroscopy: measurement of a quantity (usually light intensity in planetary astronomy) as a function of either wavelength or frequency

storm surge: increase in sea level due to winds piling up water on the right-hand leading edge of a tropical cyclone (left-hand leading edge in the southern hemisphere)

stress fracture: mechanical/structural failure under the influence of stresses

strike-slip fault: fault whose surface is typically vertical or nearly so and where the motion along the fault is parallel to the strike of the fault surface (the fault blocks move sideways past each other)

subduction: process by which one tectonic plate moves under another tectonic plate and sinks into the mantle as the plates converge

subduction zone: area where two tectonic plates move toward each other and subduction occurs

sublimation: phase change directly from solid (ice) to gas (vapor), bypassing the liquid phase

sunspot: dark, cooler region in the Sun's photosphere with a strong magnetic field

supersonic: faster than the speed that sound waves can travel in a given medium

synchronous rotation: rotation of a planet or satellite with a period equal to its orbital period

tektite: small, dark, glassy rock, spherical or elongated in shape, likely formed from rapid cooling of splashed molten rock during an asteroid/cometary impact

terminator: dark line marking the boundary between day and night on a planetary body

terrestrial planet: rocky, Earth-like planet made up predominantly of silicate rock

Tertiary period: geologic period 65–2.6 million years ago, a period cooler than the previous Cretaceous period and associated with the beginning of the Age of Mammals

theoretical model: simplified representation based on mathematical or physical arguments

theory: explanation or model that covers a substantial group of occurrences in nature and has been confirmed by a substantial number of experiments and observations

thermocline: transition in the ocean from warm mixed water near the surface to deep cold water; the depth ranges from roughly 50 m (160 ft) in the equatorial eastern Pacific to as deep as 1,000 m (0.6 mile) in midlatitudes

thermonuclear: describing nuclear fusion processes that require high temperatures to initiate

thermophile: type of extremophile organism that thrives in, or possibly requires, relatively high temperatures, e.g. 45°C–80°C (113°F–176°F)

thrust fault: break in the Earth's crust where rocks from below are pushed up and over rocks from higher strata

tidal acceleration: acceleration of a moon in its orbit about a planet due to a gravitational torque between the moon and a tidal bulge of the planet (conservation of angular momentum requires the planet to be decelerated in its rotation)

tidal bore: wave generated as incoming ocean tide is channeled through a narrow estuary or river

tidal flexing: deformation of the surface of a planet or moon due to variation in tidal forces

tidal force: secondary force caused by differences in the gravitational force from one side of an object to the other side

tidal heating: frictional heating of a satellite's interior due to flexure caused by the gravitational pull of its parent planet and possibly neighboring satellites

tidal locking: synchronous rotation of a planet or moon caused by tidal forces

topography: distribution and elevation of surface features, such as mountains, rivers, valleys, etc.

torus: doughnut-shaped ring of gas particles in the orbits of Jupiter's moons Io and Europa

transform fault: special variety of strike-slip fault that accommodates relative horizontal slip between other tectonic elements (such as oceanic crustal plates)

transit: motion of an apparently smaller celestial body across the face of an apparently larger body

Trans-Neptunian Object (TNO): any object in the Solar System that orbits the Sun at a greater distance, on average, than Neptune

Triassic period: geologic period 251–199 million years ago, generally characterized as warm and dry and the beginning of the Age of Reptiles

Trojan: body that orbits 60° ahead of or behind another body in the same orbit at the L4 and L5 Lagrange points

tropical cyclone: cyclone that develops in the tropics; called a hurricane in the Atlantic and eastern Pacific oceans, typhoon in the western Pacific, and tropical cyclone in the south Pacific and Indian oceans

tsunami: ocean wave generated by displacement of a large volume of water, for example by earthquake, volcanic eruption, bolide impact, etc.

typhoon: tropical cyclone in the northwest Pacific Ocean, west of the dateline

ultraviolet: region of the electromagnetic spectrum with wavelengths from 100 to 400 nm, shorter than that of visible light, but longer than X-rays

umbra: darkest part of a shadow caused by an occulting body in which the light source is completely obscured

uplift: geologic process through which a region is made to rise above adjacent areas

viscoelastic: exhibiting both viscous and elastic characteristics when undergoing deformation

viscosity: measure of the resistance of a fluid that is being deformed; a fluid with large viscosity is more resistant to deformation

visible light: region of the electromagnetic spectrum from 400 nm to 700 nm in wavelength

visible window: range of visible wavelengths in which very little radiation is absorbed by the atmosphere

volatile: substance that readily evaporates at normal temperatures and pressures

volcanic vent: opening in a planet's crust where molten lava and volcanic gases escape onto the surface or into the atmosphere

Walker circulation: zonal atmospheric circulation in the equatorial Pacific Ocean, normally with rising air over Indonesia and sinking air over the eastern Pacific

white dwarf: small, dense star composed of carbon and oxygen but without thermonuclear reactions

X-ray: region of the electromagnetic spectrum from 0.01 nm to 10 nm in wavelength, shorter than ultraviolet wavelengths

Acronyms

AU: astronomical unit

AUI: Associated Universities, Inc.

AURA: Association of Universities for Research in Astronomy

BCE: before the Common Era

CAMEX: Convection And Moisture EXperiment

CC: Creative Commons license

CFC: chlorofluorocarbon

CIRS: Composite Infrared Spectrometer

CME: coronal mass ejection

CRISM: Compact Reconnaissance Imaging Spectrometer for Mars

CXC: Chandra X-Ray Observatory

DLR: Deutsche Zentrum für Luft-und Raumfahrt (German Aerospace Center)

DMSP: Defense Meteorological Satellite Program

DNA: deoxyribonucleic acid

ENA: energetic neutral atom

ENSO: El Niño Southern Oscillation

ESA: European Space Agency

ESO: European Southern Observatory

FU Berlin: Freie Universität Berlin (Free University Berlin)

GFDL: GNU Free Documentation License

GRS: Great Red Spot

GSFC: Goddard Space Flight Center

HiRISE: High Resolution Imaging Science Experiment

HRSC: High Resolution Stereo Camera

HST: Hubble Space Telescope

HUP: Harvard University Press

IAU: International Astronomical Union

IBEX: Interstellar Boundary Explorer

IPCC: Intergovernmental Panel on Climate Change

ISAS: Institute of Space and Astronautical Science

JAXA: Japan Aerospace Exploration Agency

JEO: Jupiter Europa Orbiter

JGO: Jupiter Ganymede Orbiter

JHUAPL: Johns Hopkins University Applied Physics Laboratory

JPL: Jet Propulsion Laboratory

KBO: Kuiper Belt object

LASCO: Large Angle and Spectrometric Coronagraph Experiment

LDAR: Lightning Detection and Ranging

LPI: Lunar and Planetary Institute

LR: Labeled Release experiment on Viking spacecraft

MATADOR: Martian Atmosphere and Dust in the Optical and Radio

MESSENGER: MErcury Surface Space ENvironment GEochemistry and Ranging

MGDS: Marine Geoscience Data System

MGS: Mars Global Surveyor

MOC: Mars Orbiter Camera

MODIS: Moderate Resolution Imaging Spectroradiometer

MOLA: Mars Orbiter Laser Altimeter

mph: miles per hour

MRO: Mars Reconnaissance Orbiter

MSSS: Malin Space Science Systems

NAIC: National Astronomy and Ionosphere Center

NASA: National Aeronautics and Space Administration

NCEP: National Centers for Environmental Prediction

NEA: near-Earth asteroid

NEAT: Near-Earth Asteroid Tracking

NEIC: National Earthquake Information Center

NEO: near-Earth object

NGDC: National Geophysical Data Center

NIA: National Institute on Aging

NIH: National Institutes of Health

NOAA: National Oceanic and Atmospheric Administration

NRAO: National Radio Astronomy Observatory

NSF: National Science Foundation

NWS: National Weather Service

OLS: Operational Linescan System

OSIRIS: Optical, Spectroscopic, and Infrared Remote Imaging System

PAH: polycyclic aromatic hydrocarbons

PPARC: Particle Physics and Astronomy Research Council

RNA: ribonucleic acid

SDO: Scattered Disk object or Solar Dynamics Observatory

SeaWIFS: Sea-viewing Wide Field-of-view Sensor

SED: Saturn Electrostatic Discharge

SOHO: Solar and Heliospheric Observatory

SSAI: Science Systems and Applications, Inc.

SSC: Spitzer Science Center

SST: sea surface temperature

STEREO: Solar Terrestrial Relations Observatory

STS: Space Transportation System

STScI: Space Telescope Science Institute

SwRI: Southwest Research Institute

THEMIS: Thermal Emission Imaging System or Time History of Events and Macroscale Interactions during Substorms

TNO: Trans-Neptunian Object

TNT: trinitrotoluene

TRACE: Transition Region and Coronal Explorer

TRMM: Tropical Rainfall Measuring Mission

UC: University of California

UMD: University of Maryland

USAF: United States Air Force

USGS: United States Geological Survey

VIMS: Visual and Infrared Mapping Spectrometer

VIRTIS: Visible and Infrared Thermal Imaging Spectrometer

VLA: Very Large Array

VMC: Venus Monitoring Camera

WHOI: Woods Hole Oceanographic Institution

Selected Bibliography

Alvarez, Walter. *T. Rex and the Crater of Doom.* Princeton, NJ: Princeton University Press, 1997.

Angelopoulos, V., J. P. McFadden, D. Larson, C. W. Carlson, S. B. Mende, H. Frey, T. Phan, D. G. Sibeck, K.-H. Glassmeier, U. Auster, E. Donovan, I. R. Mann, I. J. Rae, C. T. Russell, A. Runov, X. Xhou, and L. Kepko. "Tail Reconnection Triggering Substorm Onset." *Science* 321 (2008):931–35. doi:10.1126/science.1160495.

Arnett, Bill. *The 9 8 Planets: A Multimedia Tour of the Solar System.* http://www.nineplanets.org.

Atreya, Sushil K. "The Mystery of Methane on Mars and Titan." *Sci Am* 296, no. 5 (2007): 42–51.

Atsma, Aaron J. *Theoi Greek Mythology: Exploring Mythology in Classical Literature & Art.* http://www.theoi.com.

Aurnou, Jonathan, Moritz Heimpel, and Johannes Wicht. "The Effects of Vigorous Mixing in a Convective Model of Zonal Flow on the Ice Giants." *Icarus* 190 (2007): 110–26.

Baalke, Ron. *Historical Background of Saturn's Rings.* http://www2.jpl.nasa.gov/saturn/back.html.

Bell, Jim. "The Red Planet's Watery Past." *Sci Am* 295, no. 6 (2006): 62–69.

Benedetti, L. R., J. H. Nguyen, W. A. Caldwell, H. Liu, M. B. Kruger, and R. Jeanloz. "Chemical Dissociation of Methane at High Pressures and Temperatures: Diamond Formation in Giant Planet Interiors?" *Science* 286 (1999): 100–102.

Black, J., P. D. Nicholson, and P. C. Thomas. "Hyperion: Rotational Dynamics." *Icarus* 72, no. 1 (1995): 149–61. doi:10.1006/icar.1995.1148.

Bullock, Mark A., and David H. Grinspoon. "Global Climate Change on Venus." *Sci Am* 280, no. 3 (1999): 50–57.

Campbell, Philip, ed. "Cassini at Jupiter." Special issue, *Nature* 415, no. 6875 (2002).

Cantor, Bruce A., Katharine M. Kanak, and Kenneth S. Edgett. "Mars Orbiter Camera Observations of Martian Dust Devils and Their Tracks (September 1997 to January 2006) and Evaluation of Theoretical Vortex Models." *J Geophys Res* 111 (2006): E12002. doi:10.1029/2006JE002700.

Canup, Robin M. "Lunar Forming Collisions with Pre-impact Rotation." *Icarus* 196 (2008): 518–38. doi:10.1016/j.icarus.2008.03.011.

Carwardine, Mark. *Extreme Nature.* New York: Collins, 2005.

Cerveny, Randy. *Freaks of the Storm: The World's Strangest True Weather Stories.* New York: Thunder's Mouth Press, 2006.

Cliver, E. W., and L. Svalgaard. "The 1859 Solar-Terrestrial Disturbance and the Current Limits of Extreme Space Weather Activity." *Solar Phys* 224 (2004): 407–22. doi:10.1007/s11207-005-4980-z.

Correia, Alexandre C. M., and Jacques Laskar. "The Four Final Rotation States of Venus." *Nature* 411 (2001): 767–70. doi:10.1038/35081000.

de Pater, Imke, and Jack J. Lissauer. *Planetary Sciences.* Cambridge, UK: Cambridge University Press, 2001.

Dicke, Ursula, and Gerard Roth. "Animal Intelligence and the Evolution of the Human Mind." *Sci Am Mind* 19, no. 4 (2008): 71–77.

Dyudina, Ulyana A., Andrew P. Ingersoll, Shawn P. Ewald, Carolyn C. Porco, Georg Fischer, William Kurth, Michael Desch, Anthony Del Genio, John Barbara, and Joseph Ferrier. "Lightning Storms on Saturn Observed by Cassini ISS and RPWS During 2004–2006." *Icarus* 190 (2007): 545–55. doi:10.1016/j.icarus.2007.03.035.

Eicher, David J., ed. *The Solar System.* Waukesha, WI: Kalmbach Publishing, 2008.

European Space Agency. *Space Science Reference: Astronomy/cosmology.* http://www.esa.int/esaSC/SEMH6QS1VED_index_0.html.

Falorini, Marco. "The Discovery of the Great Red Spot." *J Brit Astron Assoc* 97 (1987): 215–19.

Gazzaniga, Michael S. *Human: The Science Behind What Makes Us Unique.* New York: Ecco, 2008.

Ghiringhelli, L. M., C. Valeriani, E. J. Meijer, and D. Frenkel. "Local Structure of Liquid Carbon Controls Diamond Nucleation." *Phys Rev Lett* 99 (2007): 055702. doi:10.1103/PhysRevLett.99.055702.

Gierasch, P. J., A. P. Ingersoll, D. Banfield, S. P. Ewald, P. Helfenstein, A. Simon-Miller, A. Vasavada, H. H. Breneman, D. A. Senske, and the Galileo Imaging Team. "Observation of Moist Convection in Jupiter's Atmosphere." *Nature* 403 (2000): 628–30.

Golub, Leon, and Jay M. Pasachoff. *Nearest Star: The Surprising Science of Our Sun.* Cambridge, MA: Harvard University Press, 2001.

Harland, David M. *Cassini at Saturn: Huygens Results.* New York: Springer Praxis, 2007.

Harmon, J. K., P. J. Perillat, and M. A. Slade. "High-resolution Radar Imaging of Mercury's North Pole." *Icarus* 149 (2001): 1–15. doi:10.1006/icar.2000.6544.

Hartmann, William K. *A Traveler's Guide to Mars: The Mysterious Landscapes of the Red Planet.* New York: Workman, 2003.

Heimpel, Moritz, and Konstantin Kabin. "Mercury Redux." *Nature Geosci* 1 (2008): 564–66. doi:10.1038/ngeo297.

Hodge, Paul. *Higher Than Everest: An Adventurer's Guide to the Solar System.* Cambridge, UK: Cambridge University Press, 2001.

Ingersoll, A. P. "Atmospheres of the Giant Planets." In *The New Solar System,* ed. J. K. Beatty, C. C. Petersen, and A. Chaikin, 4th ed., 201–20. Cambridge, UK: Cambridge University Press and Sky Publishing Corporation, 1999.

International Astronomical Union. *Minor Planet Center.* http://www.cfa.harvard.edu/iau/mpc.html.

———. *Planet Definition Questions & Answers Sheet.* http://www.iau.org/public_press/news/release/iau0601/q_answers.

Jet Propulsion Laboratory. *Cassini Equinox Mission.* http://saturn.jpl.nasa.gov.

———. *Galileo—Journey to Jupiter.* http://www2.jpl.nasa.gov/galileo.

———. *Planetary Photojournal.* http://
photojournal.jpl.nasa.gov.

———. *Planetary Satellite Physical Parameters.*
http://ssd.jpl.nasa.gov/?sat_ phys_par.

———. *Planets and Pluto: Physical
Characteristics.* http://ssd.jpl.nasa.
gov/?planet_phys_par.

———. *Solar System Dynamics.* http://ssd.jpl
.nasa.gov/.

———. *Stardust—NASA's Comet Sample Return
Mission.* http://stardust.jpl. nasa.gov.

Jewitt, David, Alessandro Morbidelli, and
Heike Rauke. *Trans-Neptunian Objects and
Comets: Saas-Fee Advanced Courses.* Berlin:
Springer, 2008.

Joint Jupiter Science Definition Team, NASA/
ESA Study Team. *Europa Jupiter System
Mission Joint Summary Report.* NASA and
ESA, January 16, 2009.

Jönsson, K. Ingemar, Elke Rabbow, Ralph O.
Schill, Mats Harms-Ringdahl, and Petra
Rettberg. "Tardigrades Survive Exposure
to Space in Low Earth Orbit." *Current
Biology* 18 (2008): R729–R731. doi:10.1016/j.
cub.2008.06.048.

Levy, David H. *Impact Jupiter: The Crash of
Comet Shoemaker-Levy 9.* Cambridge, MA:
Basic Books, 1995.

Littmann, Mark, Fred Espenak, and Ken
Willcox. *Totality: Eclipses of the Sun.* 3rd ed.
Oxford: Oxford University Press, 2008.

Lopes, Rosaly M. C., and Michael W. Carroll.
Alien Volcanoes. Baltimore, MD: The Johns
Hopkins University Press, 2008.

Lopes, Rosaly M. C., and John R. Spencer. *Io
after Galileo: A New View of Jupiter's Volcanic
Moon.* New York: Springer/Praxis, 2007.

Lorenz, Ralph, and Jacqueline Mitton.
*Titan Unveiled: Saturn's Mysterious
Moon Explored.* Princeton, NJ: Princeton
University Press, 2008.

Lunine, Jonathan I. *Earth: Evolution of a
Habitable World.* Cambridge, UK: Cambridge
University Press, 1998.

McFadden, Lucy-Ann, Paul Weissman,
Torrence Johnson, and Linda Versteeg-
Buschman, eds. *Encyclopedia of the Solar
System.* 2nd ed. London: Academic Press,
2006.

McKeegan, K. D., A. B. Kudryavtsev, and J.
W. Schopf. "Raman and Ion Microscopic
Imagery of Graphitic Inclusions in
Apatite from Older Than 3830 Ma Akilia
Supracrustal Rocks, West Greenland."
Geology 35 (2007): 591–94. doi:10.1130/
G23465A.1.

Miller, Ron, and William K. Hartmann. *The
Grand Tour: A Traveler's Guide to the Solar
System.* New York: Workman, 2005.

NASA. *Astrobiology: Life in the Universe.* http://
astrobiology.nasa.gov.

———. *NASA Eclipse Web Site.* http://eclipse
.gsfc.nasa.gov/eclipse.html.

———. *NASA Images.* http://nasaimages.org.

———. *Solar System Exploration.* http://
solarsystem.nasa.gov/index.cfm.

———. *Solar System Exploration—Deep Impact
Legacy Site.* http://solarsystem.nasa.gov/
deepimpact.

National Oceanic and Atmospheric Administration. *Climate Prediction Center—Climate & Weather Linkage: El Niño–Southern Oscillation.* http://www.cpc.noaa.gov/products/precip/CWlink/MJO/enso.shtml.

———. *National Climatic Data Center: Extreme Weather and Climate Events.* http://www.ncdc.noaa.gov/oa/climate/severeweather/extremes.html.

———. National Hurricane Center Web Site. http://www.nhc.noaa.gov.

———. *NWS JetStream—An Online School for Weather.* http://www.srh.noaa.gov/jetstream.

Paige, David A., Stephen E. Wood, and Ashwin R. Vasavada. "The Thermal Stability of Water Ice at the Poles of Mercury." *Science* 258 (1992): 643–48. doi:10.1126/science.258.5082.643.

Pappalardo, Robert T., William B. McKinnon, and Krishan Khurana, eds. *Europa.* Tucson: University of Arizona, 2009.

The Planetary Society. *Our Solar System Space Topics.* http://www.planetary.org/explore/topics/groups/our_solar_system.

Prockter, Louise M. "Ice in the Solar System." *Johns Hopkins APL Technical Digest* 26, no. 2 (2005): 175–88. http://www.jhuapl.edu/techdigest/td2602/Prockter.pdf.

Sagan, Carl. *The Demon-Haunted World: Science As a Candle in the Dark.* New York: Ballantine, 1997.

SaveTheSea. *Save The Sea (Ocean Facts).* http://www.savethesea.org/index2.html.

Schubert, Gerald, Don L. Turcotte, and Peter Olson. *Mantle Convection in the Earth and Planets.* New York: Cambridge University Press, 2001.

Slade, Martin A., Bryan J. Butler, and Duane O. Muhleman. "Mercury Radar Imaging: Evidence for Polar Ice." *Science* 258 (1992): 635–40. doi:10.1126/science.258.5082.635.

Smith, Michael D., Barney J. Conrath, John C. Pearl, and Philip R. Christensen. "Thermal Emission Spectrometer Observations of Martian Planet-Encircling Dust Storm 2001A." *Icarus* 157 (2002): 259–63.

Solomon, Sean C., Ralph L. McNutt, Jr., Thomas R. Watter, David J. Lawrence, William C. Feldman, James W. Head, Stamatios M. Krimigis, Scott L. Murchie, Roger J. Phillips, James A. Slavin, and Maria T. Zuber. "Return to Mercury: A Global Perspective on MESSENGER's First Mercury Flyby." *Science* 321, no. 59 (2008). doi:10.1126/science.1159706.

Solomon, S., D. Qin, M. Manning, Z. Chen, M. Marquis, K. B. Averyt, M. Tignor, and H. L. Miller, eds. *Climate Change 2007: The Physical Science Basis. Contribution of Working Group I to the Fourth Assessment Report of the Inter-governmental Panel on Climate Change.* Cambridge, UK: Cambridge University Press, 2007.

Southwest Research Institute. *What Defines the Boundary of the Solar System?* http://www.ibex.swri.edu/students/What_defines_the_boundary.shtml.

Space Telescope Science Institute. *Hubblesite—Picture Album: Solar System.* http://hubblesite.org/gallery/album/solar_system.

Sparrow, Giles. *The Traveler's Guide to the Solar System.* New York: Collins, 2006.

Sromovsky, L. A., P. M. Fry, W. M. Ahue, H. B. Hammel, I. de Pater, K. A. Rages, M. R. Showalter, and M. A. van Dam. "Uranus at Equinox: Cloud Morphology and Dynamics." *Bull Am Astron Soc* 40 (2008): 488.

Stevenson, David. "Lunar Mysteries Beckon." *The Planetary Report* 27 (2007): 6–9.

Tittemore, William C., and Jack Wisdom. "Tidal Evolution of the Uranian Satellites: III. Evolution Through the Miranda-Umbriel 3:1, Miranda-Ariel 5:3, and Ariel-Umbriel 2:1 Mean-Motion Commensurabilities." *Icarus* 85, no. 2 (1990): 394–443. doi:10.1016/0019 -1035(90)90125-S.

University of Arizona. Phoenix Mars Mission Web Site. http://phoenix.lpl.arizona.edu.

U.S. Geological Survey. Astrogeology Science Center Web Site. http://astrogeology.usgs.gov.

———. *Gazetteer of Planetary Nomenclature.* http://planetarynames.wr.usgs.gov.

Vasavada, Ashwin R., and Adam P. Showman. "Jovian Atmospheric Dynamics: An Update after Galileo and Cassini." *Rep Prog Phys* 68 (2005): 1935–96. doi:10.1088/0034– 4885/68/8/R06.

Ward, Peter, and Donald Brownlee. *Rare Earth: Why Complex Life Is Uncommon in the Universe.* New York: Springer-Verlag. 2000.

Woods Hole Oceanographic Institute. *Dive and Discover: Expeditions to the Seafloor.* http:// www.divediscover.whoi.edu.

Photo Credits

Page 93   Imke de Pater (Univ. of California–Berkeley)/Heidi Hammel, Space Science Institute/Lawrence Sromovsky and Patrick Fry (Univ. of Wisconsin–Madison); obtained at the Keck Observatory, Kamuela, Hawai'i

Page 95   Hinode JAXA/NASA/PPARC

Page 96   Top: NASA; bottom left: NRAO/AUI/NSF; bottom right: NAIC—Arecibo Observatory, a facility of the NSF

Page 97   Martin Slade on behalf of the Caltech/JPL team of Dewey Muhleman, Bryan Butler, and Martin Slade, using NRAO's VLA and NASA's Goldstone Radar facilities

Page 98   John Harmon, Arecibo Observatory

Page 99   NASA/JPL/Space Science Institute

Page 101   Top: NASA/JPL/Space Science Institute; bottom: Galileo Galilei

Page 102   NASA/the Hubble Heritage Team (STScI/AURA)

Page 103   Top: NASA/JPL; bottom: NASA/JPL/Space Science Institute

Page 104   Top: NASA/JPL/Space Science Institute; bottom: NASA/JPL/Space Science Institute

Page 105   Top right: NASA/JPL/Cornell; top left: NASA/Erich Karkoschka (Univ. of Arizona); bottom: NASA/JPL

Page 106   Laura Melvin

Page 107   Philipp Salzgeber

Page 108   NASA/JPL/R. Hurt (SSC)

Page 109   NASA/JPL/Interstellar Probe Science Definition Team

Page 110   H. A. Weaver, T. E. Smith (Space Telescope Science Institute), and J. T. Trauger/R. W. Evans (JPL)/NASA

Page 111   Top: H. A. Weaver/T. E. Smith (Space Telescope Science Institute)/NASA/Laura Melvin; bottom: Peter McGregor (Australian National University); NASA/JPL

Page 112   HST Jupiter Imaging Team

Page 113   R. Evans/J. Trauger/H. Hammel/HST Comet Science Team/NASA

Page 114   NASA/JPL/T. Pyle (SSC)

Page 116   NASA/JPL

Page 117   after Alan Chamberlain (JPL/Caltech)

Page 118   Montage by Emily Lakdawalla (the Planetary Society). All images NASA/JPL/Ted Stryk except Mathilde: NASA/JHUAPL/Ted Stryk; Steins: ESA/OSIRIS team; Eros: NASA/JHUAPL; Itokawa: ISAS/JAXA/Emily Lakdawalla

Page 120   NASA/JPL/USGS

Page 121   NASA/JPL

Page 122   NASA

Page 123   NASA/JPL

Page 125   NASA/CXC/SwRI/R. Gladstone and NASA/ESA/Hubble Heritage Team

Page 127   SOHO/ESA/NASA

Page 128   Hinode JAXA/NASA/PPARC

Page 129   Top: TRACE/Stanford-Lockheed/NASA; bottom: SOHO/ESA/NASA

Page 130   SOHO/ESA/NASA

Page 131   NASA/JPL/Walt Feimer

Page 132   Top: Steele Hill/SOHO/ESA/NASA; bottom: ESA/C. Carreau

Page 133   Dr. Tony Phillips

Page 134   SwRI

Page 135   NASA/CXC/M. Weiss

Page 136 NASA/JPL/Johns Hopkins University

Page 137 NASA/JPL

Page 138 John Spencer; NASA/ESA/John T. Clarke (Univ. of Michigan)

Page 139 Joshua Strang (USAF)

Page 140 NASA, STS-39

Page 141 Laura Melvin, after NASA/CXC/M. Weiss

Page 142 NASA/CXC/SwRI/R. Gladstone and NASA/ESA/Hubble Heritage Team

Page 143 John Clarke (Univ. of Michigan)/ NASA

Page 144 NASA

Page 145 Laura Melvin, after NASA Global Hydrology and Climate Center

Page 146 NASA/JPL

Page 147 NASA/JPL/Space Science Institute

Page 148 NASA/JPL/Univ. of Iowa

Page 149 NASA

Page 151 NASA

Page 152 Laura Melvin, after IAU/Martin Kornmesser

Page 153 Laura Melvin

Page 154 Left side: UC–Berkeley Electron Microscope Lab; Patrick Edwin Moran; Richard Ling; N. Copley/ WHOI; right side: NASA; U.S. Botanical Garden; Laney Baker

Page 156 Don Davis, NASA

Page 158 Calvin Hamilton

Page 159 V. L. Sharpton, LPI

Page 160 Leonid Kulik Expedition

Page 161 Julian Baum/Take 27 Ltd.

Page 162 Laura Melvin

Page 163 NASA/JPL/UMD; NASA/JPL/R. Hurt (SSC)

Page 164 NASA/JPL

Page 166 Tom Ruen/Eugene Antoniadi/Lowell Hess/Roy A. Gallant/HST/NASA

Page 167 Top: Stanley Sheff; bottom: NASA/ JPL/Roel van der Hoorn

Page 168 NASA

Page 169 NASA/JPL/JHUAPL/MSSS/Brown University

Page 170 V. Tunnicliffe (Univ. of Victoria)

Page 171 NOAA Ocean Explorer

Page 172 Nicolle Rager Fuller (NSF)

Page 173 S. Hengherr/R. Schill/K. H. Hellmer

Page 174 NASA/JPL/Michael Carroll

Page 175 NASA

Page 177 NASA/JPL/Univ. of Arizona

Page 178 NASA/JPL/Lowell Observatory

Page 179 Top: NASA/JHUAPL/SwRI; bottom: NASA/JPL/Univ. of Arizona

Page 180 NASA/JPL/Arizona State University

Page 181 NASA/JPL/Space Science Institute

Page 182 Top: NASA/JPL/Laura Melvin; bottom: NASA/JPL/ESA/Univ. of Arizona

Page 183 NASA/JPL/Space Science Institute

Page 184 NASA/JPL/USGS

Page 185 NASA/ ESA/H. Weaver (JHUAPL)/A. Stern (SwRI)/the HST Pluto Companion Search Team

Page 186 International Astronomical Union

Page 187 NASA

Page 188 NASA/ESA/A. Feild (STScI)

Page 190 Laura Melvin, after images by Calvin Hamilton

Page 241 NASA
Page 242 D. Baker
Page 243 M. Ruzek

Page 244 Data courtesy Marc Imhoff of NASA
          GSFC and Christopher Elvidge
          of NOAA NGDC; image by Craig
          Mayhew and Robert Simmon of
          NASA GSFC

Acknowledgments

When we set out to write this book, we had no idea what lay in store for us. Fortunately, several individuals and institutions were of tremendous help along the way.

First and foremost, we would like to thank the scientists, engineers, policymakers, and support staff that make the dream of Solar System exploration a reality. Space exploration programs and interplanetary missions involving many international partners have released to the world some truly amazing discoveries. Without their hard work, we wouldn't have nearly so many cool things to discuss in this book.

Our agents guided, cajoled, and encouraged until the book idea became a proposal, the proposal became a deal, and the deal became a manuscript. Babette Sparr kept the project on track from start to finish. Her expertise and enthusiasm guided us through the surreal experience of signing our first book deal and landed us our first translated version (German).

Thanks go as well to the editors and staff of Harvard University Press and Rowohlt Verlag. Frank Strickstrock at Rowohlt was immediately enthusiastic about the project and was eager to see the text translated into German. Michael Fisher took a chance on two first-time authors who had something a little different from what HUP usually publishes. We hope they enjoy how it all turned out.

Several people gave generously of their time by reading and commenting on drafts of early chapters: Scott Curtis, Larry Appleby, Michael Cleveland, Tom Baker, Heather Quantz, Josh Kaplan, Keri Jones, Joseph Chaves, Parth Shah, and Isabel Hernandez. Their feedback helped refine our vision for the scope of the book and also helped set the tone for our subsequent writing. Two anonymous reviewers provided in-depth and thoughtful critiques of the completed first draft. Incorporating their suggestions made us think and led to an improved final book.

In addition to the freely available images from the major space programs, many individuals graciously allowed us to reproduce images of their own creation within these pages. Although there are too many to name individually, we are very grateful for their willingness to share their work. When we just couldn't find the exact images we needed to get our point across, we turned to our graphic artist, Laura Melvin. She managed to produce some lovely illustrations despite having to work with a couple of picky scientists. Pam Kabir helped us wade through the morass of getting permission to use all of the 275-plus images in the book.

Austin College students helped guide the development of the book, sometimes unknowingly during impromptu discussions of extreme phenomena in class. Nathan Drake, Matt Varvir, Jordan Robison, David Riddle, and Lauren Dorsett conducted research to support what we were writing. Thomas Joiner took a few images and a small amount of text from us and turned them into a web site that exceeded our expectations. David Baker was supported during this project by the Austin College Richardson Grant.

And, of course, without the extreme patience, extreme support, and extreme understanding of our families, *The 50 Most Extreme Places in Our Solar System* would never have made it past the idea phase and onto paper. We are lucky fellows.

Finally, we'd like to thank the Solar System for being so extreme.

Index